Essentials of Monte Carlo Simulation

Nick T. Thomopoulos

Essentials of Monte Carlo Simulation

Statistical Methods for Building Simulation Models

 Springer

Nick T. Thomopoulos
Stuart School of Business
Illinois Institute of Technology
Chicago, Illinois, USA

ISBN 978-1-4614-6021-3   ISBN 978-1-4614-6022-0 (eBook)
DOI 10.1007/978-1-4614-6022-0
Springer New York Heidelberg Dordrecht London

Library of Congress Control Number: 2012953256

Printed on acid-free paper

Springer is part of Springer Science+Business Media (www.springer.com)

For my wife,
my children,
and my grandchildren.

Preface

I was fortunate to have a diverse career in industry and academia. This included working at International Harvester as supervisor of operations research in the corporate headquarters; at IIT Research Institute (IITRI) as a senior scientist with applications that spanned worldwide in industry and government; as a professor in the Industrial Engineering Department at the Illinois Institute of Technology (IIT), in the Stuart School of Business at IIT; at FIC Inc. as a consultant for a software house that specializes in supply chain applications; and the many years of consulting assignments with industry and government throughout the world. At IIT, I was fortunate to be assigned a broad array of courses, gaining a wide breadth with the variety of topics, and with the added knowledge I acquired from the students, and with every repeat of the course. I also was privileged to serve as the advisor to many bright Ph.D. students as they carried on their dissertation research. Bits of knowledge from the various courses and research helped me in the classroom, and also in my consulting assignments. I used my industry knowledge in classroom lectures so the students could see how some of the textbook methodologies actually are applied in industry. At the same time, the knowledge from the classroom helped to formulate and develop Monte Carlo solutions to industry applications as they unfolded. This variety of experience allowed the author to view how simulation can be used in industry. This book is based on this total experience.

Simulation has been a valuable tool in my professional life, and some of the applications are listed below. The simulations models were from real applications and were coded in various languages of FORTRAN, C++, Basic, and Visual Basic. Some models were coded in an hour, others in several hours, and some in many days, depending on the complexity of the system under study. The knowledge gained from the output of the simulation models proved to be invaluable to the research team and to the project that was in study. The simulation results allowed the team to confidently make the decisions needed for the applications at hand. For convenience, the models below are listed by type of application.

Time Series Forecasting

- Compare the accuracy of the horizontal forecast model when using 12, 24 or 36 months of history.
- Compare the accuracy of the trend forecast model when using 12, 24 or 36 months of history.
- Compare the accuracy of the seasonal forecast model when using 12, 24, or 36 months of history.
- Compare the accuracy of forecasts between weekly and monthly forecast intervals.
- Compare the accuracy benefit of forecasts when using month-to-date demands to revise monthly forecasts.
- Compare the accuracy of the horizontal forecast model with the choice of the alternative forecast parameters.
- Compare the accuracy of the trend forecast model with the choice of the alternative forecast parameters.
- Compare the accuracy of the seasonal forecast model with the choice of the alternative forecast parameters.
- In seasonal forecast models, measure how the size of the forecast error varies as the season changes from low-demand months to high-demand months.

Order Quantity

- Compare the inventory costs for parts (with horizontal, trend, and seasonal demand patterns) when stock is replenished by use of the following strategies: EOQ, Month-in-Buy or Least Unit Cost.
- Compare various strategies to determine the mix of shoe styles to have in a store that yields the desired service level and satisfies the store quota.
- Compare various strategies to determine the mix of shoe sizes for each style type to have in a store that yields the desired service level and satisfies the store quota for the style.
- Compare various strategies to find the initial-order-quantity that yields the least cost for a new part in a service parts distribution center.
- Compare various strategies to find the all-time-requirement that yields the least cost for a part in a service parts distribution center.
- Compare various ways to determine how to measure lost sales demand for an individual part in a dealer.
- Compare strategies, for a multi-distribution system, on how often to run a transfer routine that determines for each part when and how much stock to transfer from one location to another to avoid mal-distribution.

Safety Stock

- Compare the costs between the four basic methods of generating safety stock: month's supply, availability, service level and Lagrange.
- Compare how the service level for a part reacts as the size of the safety stock and the order quantity vary.
- Compare how a late delivery of stock by the supplier affects the service level of a part.
- Compare strategies on how to find the minimum amount of early stock to have available to offset the potential of late delivery by the supplier.
- Measure the relationship between the service level of a part and the amount of lost sales on the part.

Production

- In mixed-model (make-to-stock) assembly, compare various strategies on how to sequence the models down the line.
- In mixed-model (make-to-order) assembly, compare various strategies on how to sequence the individual jobs down the line.
- In job-shop operations, determine how many units to initially produce to satisfy the order needs and minimize the material, machine, and labor costs.
- In machine-loading operations, compare strategies on how to schedule the jobs through the shop to meet due dates and minimize machine idle times.
- Compare strategies on how to set the number of bays (for maintenance and repair) in a truck dealership that meets the customer needs and minimizes the dealer labor costs.

Other

- In the bivariate normal distribution, estimate the cumulative distribution function for any combination of observations when the means and variances are given, and the correlation varies between -1.0 and 1.0.
- In the bivariate lognormal distribution, estimate the cumulative distribution function for any combination of observations when the means and variances of the transformed variables are known, and the correlation varies from -1.0 to 1.0.
- In the multivariate normal distribution with k variables, an estimate of the cumulative distribution function is obtained, for any combination of observations when both the mean vector and the variance-covariance matrix are known.

- In an airport noise abatement study, noise measures were estimated, as in a contour map, for the airport and for all blocks surrounding the airport. The noise was measured with various combinations of: daily number of flights in and out, the type of aircraft and engines, and the direction of the runways in use.
- In a study for the navy, some very complex queuing systems were in consideration. Analytical solutions were developed, and when a level of doubt was present in the solution, simulation models were developed to verify the accuracy of the analytical solutions.
- A simulated database for part numbers were needed in the process of developing various routines in forecasting and inventory replenishment for software systems. These were for systems with one or more stocking locations. The database was essential to test the effectiveness of the routines in carrying out its functions in forecasting and inventory replenishments. The reader may note that many of the fields in the database were jointly related and thereby simulated in a correlated way.

Acknowledgments

Thanks especially to my wife, Elaine Thomopoulos, who encouraged me to write this book, and who gave consultation whenever needed. Thanks also to the many people who have helped and inspired me over the years, some of whom are former IIT students from my simulation classes. I can name only a few here: Bob Allen (R. R. Donnelly), Wayne Bancroft (Walgreens), Fred Bock (IIT Research Institute), Harry Bock (Florsheim Shoe Company), Dan Cahill (International Truck and Engine), Debbie Cernauskas (Benedictine University), Dick Chiapetta (Chiapetta, Welch and Associates), Edine Dahel (Monterey Institute), Frank Donahue (Navistar), John Garofalakis (Patras University), Tom Galvin (Northern Illinois University), Tom Georginis (Lewis University), Shail Godambe (Motorola, Northern Illinois University), M. Zia Hassan (Illinois Institute of Technology), Willard Huson (Navistar), Robert Janc (IIT Research Institute), Marsha Jance (Indiana University – Richmond), Chuck Jones (Decision Sciences, Inc.), Tom Knowles (Illinois Institute of Technology), Joachim Lauer (Northern Illinois University), Carol Lindee (Panduit), Anatol Longinow (IIT Research Institute), Louca Loucas (University of Cyprus), Nick Malham (FIC Inc.), Barry Marks (IIT Research Institute), Jamshid Mohammadi (Illinois Institute of Technology), Fotis Mouzakis (Cass Business School of London), Pricha Pantumsinchai (M-Focus), Ted Prenting (Marist College), Ornlatcha Sivarak (Mahidol University), Spencer Smith (Illinois Institute of Technology), Mark Spieglan (FIC Inc), Paul Spirakis (Patras University) and Tolis Xanthopoulos (IIT).

Nick T. Thomopoulos

Contents

Chapter 1
Introduction

Monte Carlo Method

To apply the Monte Carol method, the analyst constructs a mathematical model that simulates a real system. A large number of random sampling of the model is applied yielding a large number of random samples of output results from the model. The origin began in the 1940s by three scientists, John von Neumann, Stanislaw Ulam and Nicholas Metropolis who were employed on a secret assignment in the Los Alamos National Laboratory, while working on a nuclear weapon project called the Manhattan Project. They conceived of a new mathematical method that would become known as the Monte Carlo method. Stanislaw Ulam coined the name after the Monte Carlo Casinos, located in Monaco. Monaco is a tiny country located just south of France facing the Mediterranean Sea, and is famous for its beauty, casinos, beaches, and auto racing. The Manhattan team formulated a model of a system they were studying that included input variables, and a series of algorithms that were too complicated to analytically solve.

The method is based on running the model many times as in random sampling. For each sample, random variates are generated on each input variable; computations are run through the model yielding random outcomes on each output variable. Since each input is random, the outcomes are random. In the same way, they generated thousands of such samples and achieved thousands of outcomes for each output variable. In order to carryout this method, a large stream of random numbers were needed. Von Neumann developed a way to calculate pseudo random numbers by using a middle-square method. Von Neumann realized the method had faults, but he reasoned the method was the fastest that was then available, and he would be aware when the method would fall out of alignment.

The Monte Carlo method proved to be successful and was an important instrument in the Manhattan Project. After the War, during the 1940s, the method was continually in use and became a prominent tool in the development of the hydrogen bomb. The Rand Corporation and the U.S. Air Force were two of the top

N.T. Thomopoulos, *Essentials of Monte Carlo Simulation: Statistical Methods for Building Simulation Models*, DOI 10.1007/978-1-4614-6022-0_1,
© Springer Science+Business Media New York 2013

organizations that were funding and circulating information on the use of the Monte Carlo method. Soon, applications started popping up in all sorts of situations in business, engineering, science and finance.

Random Number Generators

A random number generator is a computerized or physical method that produces numbers that have no sequential pattern and are arranged purely by chance. Since the early times, many ways have been applied to generate random deviates, like: rolling dice, flipping coins, roulette wheels, and shuffling cards. These physical methods are not practical when a large number of random numbers are needed in applications. In 1947, the Rand Corporation generated random numbers by use of an electronic roulette type device that was connected to a computer. A book, with a list of all the numbers, was published by the Rand Corporation (1946). The numbers were also available on punched cards and tape. The numbers had found many applications in statistics, experimental design, cryptography and other scientific disciplines. However, with the advent of high-speed computers in the 1950s, mathematical algorithms became practical and new developments led to improved ways of generating a large stream of random numbers.

Computer Languages

Since the 1940s, many computer languages have been developed and in use in one way or another allowing programmers to write code for countless applications. Early languages like: COBOL, FORTRAN, Basic, Visual Basic , JAVA and C++, were popular in developing computer simulation models.

As simulation became more popular in solving complex problems in business, engineering, science, and finance, a new set of computer languages (e.g., GAUSS, SAS, SPSS, R) have evolved. Some are for either of the two major types of simulation: continuous and discrete, and some can handle both. These languages allow the user to construct the process he/she is emulating and also has the ability to gather statistics, perform data analysis and provide tables and graphs on outcome summaries.

Discrete simulation models are when the events change one at a time, like in queuing models where new customers arrive or depart the system individually or in batches. Continuous simulation models are when the events are continuously changing over time, according to a set of differential equations. Could be the trajectory of a rocket.

Computer Simulation Software

The number of simulation software packages has also exploded over the years and most apply for specific processes in industry, gaming and finance. The software may include a series of mathematical equations and algorithms that are associated with a given process. When the software fits the process under study, the user can apply the software and quickly observe the outcomes from a new or modified arrangement of the process without actually developing the real system. This ability is a large savings in time and cost of development.

In the 1990s, Frontline presented the Solver application that was for spreadsheet use to solve linear and nonlinear problems. This soon led to Microsoft's Excel Solver. Improvements were made and in the 2000s, Monte Carlo Simulation was introduced as another Excel application where many trials of a process are automatically performed from probability distributions specified by the user. In 2006, yet another added feature is the RISK Solver Engine for EXCEL that performs instant Monte Carlo Simulations whenever a user changes a number on a spreadsheet.

Basic Fundamentals

The early pioneers of the Manhattan Project were fully aware that the validity of their model highly depended on the authenticity of the algorithms they formed as well as the choice of the input probability distributions and parameter values they selected. An error in the formulation could give misleading results. With their creativity, intellect and careful construction, the output from their simulation model was highly successful.

Monte Carlo methods are now extensively used in all industries and government to study the behavior of complex systems of all sorts. Many of the applications are performed with software programs, like the Excel Solver models, described earlier. The casual user will run the powerful easy-to-use software model and collect the results, and with them, make decisions in the work place, and that might be all that is needed. This user may not be completely aware of how the model works inside and may not have a need to know.

Another user may want to know more on how the model does what it does. Many others who code their own simulation models need to know the fundamentals. This book is meant for these users, giving the mathematical basis on developing simulation models. A description is given on how pseudo random numbers are generated, and further, how they are used to generate random variates of input variables that come from specified probability distributions, discrete or continuous.

The book further describes how to cope with simulation models that are associated with two or more variables that are correlated and jointly related. These are called multivariate variables, and various distributions of this type are described. Some are discrete, like the multinomial, multivariate hyper geometric, and some are continuous like the multivariate normal and multivariate lognormal.

In addition, the text helps those users who are confronted with a probability distribution that does not comply with those that are available in the software in use. Further, the system could be a non-terminating system that includes transient and equilibrium (steady state) stages. The text also gives insight on how to determine the end of the transient stage and the beginning of the equilibrium stage. Most analysts only want to collect and analyze the data from the equilibrium stage.

Further, one chapter shows how to generate output data that are independent so they can properly be analyzed with statistical methods. Methods are described to steer the results so that the output data is independent. Another chapter presents a review on the common statistical methods that are used to analyze the output results. This includes the measuring of the average, variance, confidence intervals, tests between two means or between two proportions, and the one-way analysis of variance.

The better the analyst can structure a simulation model to emulate the real system, the more reliable the output results in problem solving decisions. Besides formulating the equations and algorithms of the system properly, the analyst is confronted with selecting the probability distributions that apply for each input variable in the model. This is done with use of the data, empirical or sample, that is available. With this data, the probability distributions are selected and the accompanying parameter values are estimated. The better the fit, the better the model. One of the chapters describes how to do this.

Another issue that sometimes confronts the analyst is to choose the probability distribution and the corresponding parameter value(s) when no data is available for an input variable. In this situation, the analyst relies of the best judgment of one or more experts. Statistical ways are offered in the text to assist in choosing the probability distribution and estimating the parameters.

Chapter Summaries

The following is a list of the remaining chapters and a quick summary on the content of each.

Chapter 2. *Random Number Generators* Since early days, the many applications of randomness have led to a wide variety of methods for generating random data of various type, like rolling dice, flipping coins and shuffling cards. But these methods are physical and are not practical when a large number of random data is needed in an application. Since the advent of computers, a variety of computational methods have been suggested to generate the random data, usually with random numbers. Scientists, engineers and researchers are ever more developing simulation models in their applications; and their models require a large – if not vast – number of random numbers in processing. Developing these simulation models is not possible without a reliable way to generate random numbers.

Chapter 3. *Generating Random Variates* Random variables are classified as discrete or continuous. Discrete is when the variable can take on a specified list of values, and continuous is when the variable can assume any value in a specified interval. The mathematical function that relates the values of the random variable with a probability is the probability distribution. When a value of the variable is randomly chosen according to the probability distribution, it is called a random variate. This chapter describes the common methods to generate random variates for random variables from various probability distributions. Two methods are in general use for this purpose, one is called the *Inverse Transform* method (IT), and the other is the *Accept-Reject* method (AR). The IT method is generally preferred assuming the distribution function transforms readily. If the distribution is mathematically complicated and not easily transformed, the IT method becomes complicated and is not easily used. The AR method generally requires more steps than the IT method to achieve the random variate. The chapter presents various adaptations of these two methods.

Chapter 4. *Generating Continuous Random Variates* A continuous random variable has a mathematical function that defines the relative likelihood that any value in a defined interval will occur by chance. The mathematical function is called the probability density. For example, the interval could be all values from 10 to 50, or might be all values zero or larger, and so forth. This chapter considers the more common continuous probability distributions and shows how to generate random variates for each. The probability distributions described here are the following: the continuous uniform, exponential, Erlang, gamma, beta, Weibull, normal, lognormal, chi-square, student's t, and Fishers F. Because the standard normal distribution is so useful in statistics and in simulation, and no closed-form formula is available, the chapter also lists the Hastings approximation formula that measures the relationship between the variable value and its associated cumulative probability.

Chapter 5. *Generating Discrete Random Variates* A discrete random variable includes a specified list of exact values where each is assigned a probability of occurring by chance. The variable can take on a particular set of discrete events, like tossing a coin (head or tail), or rolling a die (1,2,3,4,5,6). This chapter considers the more common discrete probability distributions and shows how to generate random variates for each. The probability distributions described here are the following: discrete arbitrary, discrete uniform, Bernoulli, binomial, hyper geometric, geometric, Pascal and Poisson.

Chapter 6. *Generating Multivariate Random Variates* When two or more random variables are jointly related in a probability way, they are labeled as multivariate random variables. The probability of the variables occurring together is defined by a joint probability distribution. In most situations, all of the variables included in the distribution are continuous or all are discrete; and on less situations, they are a mixture between continuous and discrete. This chapter considers some of the more popular multivariate distributions and shows how to generate random variates for each. The probability distributions described here are the following: multivariate discrete arbitrary, multinomial, multivariate hyper geometric, bivariate normal, bivariate lognormal, multivariate normal and multivariate lognormal. The Cholesky decomposition method is also described since it is needed to generate random variates from the multivariate normal and the multivariate lognormal distributions.

Chapter 7. *Special Applications* This chapter shows how to generate random variates for applications that are not directly bound by a probability distribution as was described in some of the earlier chapters. The applications are instructively useful and often are needed as such in simulation models. They are the following: Poisson process, constant Poisson process, batch arrivals, active redundancy, standby redundancy, random integers without replacement and poker.

Chapter 8. *Output From Simulation Runs* Computer simulation models are generally developed to study the performance of a system that is too complicated for analytical solutions. The usual goal of the analyst is to develop a computer simulation model that emulates the activities of the actual system as best as possible. Many of these models are from terminating and nonterminating systems.

A terminating system is when a defined starting event B and an ending event C are specified, and so, each run of the simulation model begins at B and ends at C. This could be a model of a car wash that opens each day at 6 a.m. and closes at 8 p.m. Each simulation run would randomly emulate the activities from B to C.

A nonterminating system is where there is no beginning or ending events to the system. The system often begins in a transient stage and eventually falls into either an equilibrium stage or a cyclical stage. This could be a study of a maintenance and repair shop that is always open. At the outset of the simulation model run, the system is empty and may take some time to enter either an equilibrium stage or a cyclical stage. This initial time period is called the transient stage.

A nonterminating system with transient and equilibrium stages might be a system where the inter-arrival flow of new customers to the shop is steadily coming from the same probability distribution. In the run of the simulation model, the system begins in the transient stage and thereafter the flow of activities continues in the equilibrium stage.

A nonterminating model with transient and cyclical stages could be a model of a system where the probability distribution of the inter-arrival flow of new customers varies by the hour of the day. The simulation run begins in a transient stage and passes to the cyclical stage thereafter.

In either system, while the analyst is developing the computer model, he/she includes code in the model to collect data of interest for later analysis. This output data is used subsequently to statistically analyze the performance of the system.

Chapter 9. *Analysis of Output Data* This chapter is a quick review on some of the common statistical tests that are useful in analyzing the output data from runs of a computer simulation model. This pertains when each run of the model yields a group of k unique output measures that are of interest to the analyst. When the model is run n times, each with a different string of continuous uniform $u \sim U(0,1)$ random variates, the output data is generated independently from run to run, and therefore the data can be analyzed using ordinary statistical methods. Some of the output data may be of the variable type and some may be of the proportion type.

The appropriate statistical method for each type of data is applied as needed. This includes, measuring the average value and computing the confidence interval of the true mean. Oftentimes, the simulation model is run with one or more control variables in a 'what if' manner. The output data between the two or more settings of the control variables can be compared using appropriate statistical tools. This includes testing for significant difference between two means, between two proportions, and between k or more means.

Chapter 10. *Choosing the Probability Distribution From Sample Data* In building a simulation model, the analyst often includes several input variables of the control and random type. The control variables are those that are of the "what if" type. Often, the purpose of the simulation model is to determine how to set the control variables in the real system seeking optimal results. For example, in an inventory simulation model, the control variables may be the service level and the holding rate, both of which are controlled by the inventory manager. On each run of the model, the analyst sets the values of the control variables and observes the output measures to see how the system reacts.

Another type of variable is the random input variables, and these are of the continuous and discrete type. This type of variable is needed to match, as best as possible, the real life system for which the simulation model is seeking to emulate. For each such variable, the analyst is confronted with choosing the probability distribution to apply and the parameter value(s) to use. Often empirical or sample data is available to assist in choosing the distribution to apply and in estimating the associated parameter values. Sometimes two or more distributions may seem appropriate and the one to select is needed. The authenticity of the simulation model largely depends on how well the analyst emulates the real system. Choosing the random variables and their parameter values is vital in this process.

This chapter gives guidance on the steps to find the probability distribution to use in the simulation model and how to estimate the parameter values that pertain. For each of the random variables in the simulation model with data available, the following steps are described: verify the data is independent, compute various statistical measures, choose the candidate probability distributions, estimate the parameter(s) for each probability distribution, and determine the adequacy of the fit.

Chapter 11. *Choosing the Probability Distribution When No Data* Sometimes the analyst has no data to measure the parameters on one or more of the input variables in a simulation model. When this occurs, the analyst is limited to a few distributions where the parameters may be estimated without empirical or sample data. Instead of data, experts are consulted who give their judgment on various parameters of the distributions. This chapter explores some of the more common distributions where such expert opinions are useful. The distributions described here are continuous and are the following: continuous uniform, triangular, beta, lognormal and Weibull. The type of data provided by the experts is the following type: minimum value, maximum value, most likely value, average value, and a p-quantile value.

Chapter 2
Random Number Generators

Introduction

For many past years, numerous applications of randomness have led to a wide variety of methods for generating random data of various type, like rolling dice, flipping coins and shuffling cards. But these methods are physical and are not practical when a large number of random data is needed in an application. Since the advent of computers, a variety of computational methods have been suggested to generate the random data, usually with random numbers. Scientists, engineers and researchers are ever more developing simulation models in their applications; and their models require a large – if not vast – number of random numbers in processing. Developing these simulation models is not possible without a reliable way to generate random numbers. This chapter describes some of the fundamental considerations in this process.

Modular Arithmetic

Generating random numbers with use of a computer is not easy. Many mathematicians have grappled with the task and only a few acceptable algorithms have been found. One of the tools used to generate random numbers is by way of the mathematical function called modular arithmetic. For a variable w, the modulo of w with modulus m is denoted as: w modulo(m). The function returns the remainder of w when divided by m. In the context here, w and m are integers, and the function returns the remainder that also is an integer. For example, if $m = 5$, and $w = 1$,

$$w \text{ modulo } (m) = 1 \text{ modulo}(5) = 1.$$

N.T. Thomopoulos, *Essentials of Monte Carlo Simulation: Statistical Methods for Building Simulation Models*, DOI 10.1007/978-1-4614-6022-0_2,
© Springer Science+Business Media New York 2013

In the same way, should w = 5, 6 or 17, then,

$$5 \bmod(5) = 0$$
$$6 \bmod(5) = 1$$
$$17 \bmod(5) = 2$$

and so forth. Hence, for example, the numbers 1, 6, 11, 16 are all congruent modulo 1 when m = 5. Note, the difference of any of the numbers, that have the same remainder, is perfectly divisible by m, and thus they are congruent. Also notice, when the parameter is m, the values returned are all integers, 0 to m-1.

An example of modular arithmetic is somewhat like the clock where the numbers for the hours are always from 1 to 12. The same applies with the days of the week (1–7), and the months of the year (1–12).

Linear Congruent Generators

In 1951, Lehmer introduced a way to generated random numbers, called the linear congruent generator, LCG. This method adapts well to computer applications and today is the most common technique in use. The method requires the use of the modulo function as shown below.

LCG calls for three parameters (a, b, m) and uses modular arithmetic. To obtain the i-th value of w, the function uses the prior value of w as in the method shown below:

$$w_i = (a \, w_{i-1} + b) \, \bmod(m)$$

In the above, w is a sequence of integers that are generated from this function, and i is the index of the sequence.

Example 2.1 Suppose, m = 32, a = 5 and b = 3. Also, assume the seed (at i = 0) for the LCG is $w_0 = 11$. Applying,

$$w_i = (5w_{i-1} + 3) \, \bmod(32)$$

yields the sequence:
26,5,28,15,14,9,16,19,2,13,4,23,22,17,24,27,10,21,12,31,30,25,0,3,18,29,20,7, 6,1,8,11

Notice there are 32 values of w for i ranging from 1 to 32, and all are different. This is called a full cycle of all the possible remainders 0–31 when m = 32. The last number in the sequence is the same as the seed, $w_0 = 11$. The sequence of numbers that are generated for the second cycle would be the same as the sequence from the first cycle of 32 numbers because the seed has been repeated.

The reader should be aware that it is not easy to find the combination of a and b that gives a full cycle for a modulus m. The trio of m $= 32$, a $= 5$ and b $= 3$ is one such combination that works.

Example 2.2 Now suppose the parameters are, m $= 32$, a $= 7$ and b $= 13$. Also assume the seed for the LCG is $w_0 = 20$. Applying,

$$w_i = (7w_{i-1} + 13) \text{ modulo}(32)$$

yields the sequence of 32 numbers:

$$25, 28, 17, 4, 9, 12, 1, 20,$$
$$25, 28, 17, 4, 9, 12, 1, 20,$$
$$25, 28, 17, 4, 9, 12, 1, 20,$$
$$25, 28, 17, 4, 9, 12, 1, 20,$$

Note there is a cycle of eight numbers, 25–20. In general, when one of the numbers in the sequence is the same as the seed, $w_0 = 20$, the sequence repeats with the same numbers, 25–20. In this situation, after eight numbers, the seed value is generated, and thereby the cycle of eight numbers will continually repeat, in a loop, as shown in the example above.

Generating Uniform Variates

The standard continuous uniform random variable, denoted as u, is a variable that is equally likely to fall anywhere in the range from 0 to 1. The LCG is used to convert the values of w to u by dividing w over m, i.e., $u = w/m$. The generated values of u will range from 0/m to $(m - 1)/m$, or from zero to just less than 1. To illustrate, the 32 values of w generated in Example 2.1 are used for this purpose. In Example 2.3, $u = w/m$ for the values listed earlier.

Example 2.3 The 32 values of u listed below (in 3 decimals) are derived from the corresponding 32 values of w listed in Example 2.1. Note, $u_i = w_i/32$ for i $= 1$–32.
    0.812, 0.156, 0.875, 0.468, 0.437, 0.281, 0.500, 0.593, 0.062, 0.406, 0.125, 0.718, 0.687, 0.531, 0.750, 0.843, 0.312, 0.656, 0.375, 0.968, 0.937, 0.781, 0.000, 0.093, 0.562, 0.906, 0.625, 0.218, 0.187, 0.031, 0.250, 0.343

32-Bit Word Length

The majority of computers today have word lengths of 32 bits. For these machines, the largest number that is recognized is $(2^{31} - 1)$, and the smallest number is $-(2^{31}-1)$. The first of the 32 bits is used to identify whether the number is a plus or a minus, leaving the remaining 31 bits to determine the number.

So, for these machines, the ideal value of the modulus is m $= (2^{31}-1)$ since a full cycle with m gives a sequence with the largest number of unique random uniform variables The goal is to find a combination of parameters that are compatible with the modulus. Fishman and Moore (1982), have done extensive analysis on random number generators determining their acceptability for simulation use. In 1969, Lewis, Goodman and Miller suggested the parameter values of a $= 16,807$ and b $= 0$; and also in 1969, Payne, Rabung and Bogyo offered a $= 630,360,016$ with b $= 0$. These combinations have been installed in computer compilers and are accepted as parameter combinations that are acceptable for scientific use. The Fishman and Moore reference also identifies other multipliers that achieve good results.

Random Number Generator Tests

Mathematicians have developed a series of tests to evaluate how good a sequence of uniform variates are with respect to truly random uniform variates. Some of the tests are described below.

Length of the Cycle

The first consideration is how many variates are generated before the cycle repeats. The most important rule is to have a full cycle with the length of the modulus m. Assume that n uniform variates are generated and are labeled as $(u_1, \ldots, u_n)$ where n $=$ m or is close to m. The ideal would be to generate random numbers where the numbers span a full cycle or very near a full cycle.

Mean and Variance

For the set of n variates $(u_1, \ldots, u_n)$, the sample average and variance are computed and labeled as $\bar{u}$ and s_u^2, respectively. The goal of the random number generator is to emulate the standard continuous uniform random variable u, denoted here as u $\sim$ U(0,1), with expected value E(u) $= 0.5$ and the variance V(u) $= \sigma^2 = 1/12 = 0.0833$. So, the appropriate hypothesis mean and variance tests are used to compare $\bar{u}$ to 0.5 and s_u^2 to 0.0833.

Chi Square

The sequence $(u_1, \ldots, u_n)$ are set in k intervals, say k $= 10$, where i $= 1$–10 identifies the interval for which each u falls. When k $= 10$, the intervals are: (0.0–0.1), (0.1–0.2), ..., (0.9–1.0). Now let f_i designate the number of u's that fall in interval i. Since n is the number of u's in the total sample, the expected number of u's in an interval is $e_i = 0.1n$. With the ten sets of f_i and e_i, a Chi Square (goodness-of-fit) test is used to determine if the sequence of u's are spread equally in the range from zero to one.

The above chi square test can be expanded to two dimensions where the pair of u's (u_i, u_{i+1}) are applied as follows. Assume the same ten intervals are used as above for both u_i and for u_{i+1}. That means there are $10 \times 10 = 100$ possible cells where the pair can fall. Let f_{ij} designate the number of pairs that fall in the cell ij. Since n values of u are tested, there are n/2 pairs. So the expected number of units to fall in a cell is $e_{ij} = 0.01n/2$. This allows use of the Chi Square test to determine if f_{ij} is significantly different than e_{ij}. For a truly uniform distribution, the number of entries in a cell should be equally distributed. When more precision is called, the length of the intervals can be reduced from 0.10 to 0.05 or to 0.01, for example.

In the same way, the triplet of u's can be tested to determine if the u's generated follow the expected values from a truly uniform distribution. With $k = 10$ and with three dimensions, the entries fall into $10 \times 10 \times 10 = 1,000$ cubes, and for a truly uniform distribution, the number of entries in the cubes should be equally distributed.

Autocorrelation

Another test computes the autocorrelation between the u's with various lags of length 1, 2, 3,….. The ideal is for all the lag autocorrelations to be significantly close to zero, plus or minus. When the lag is k, the estimate of the autocorrelation is the following,

$$r_k = \sum_i (u_i - 0.5)(u_{i-k} - 0.5) / \sum_i (u_i - 0.5)^2$$

Pseudo Random Numbers

In the earlier example when $m = 32$, a full cycle of w's were generated with the parameters $(a = 5, b = 3)$. Further when the seed was set at $w_0 = 11$, the sequence of w's generated were the following:

26,5,28,15,14,9,16,19,2,13,4,23,22,17,24,27,10,21,12,31,30,25,0,3,18,29,20,7, 6,1,8,11

With this combination of parameters, whenever the seed is $w_0 = 11$, the same sequence of random numbers will emerge. So in essence, these are not truly random numbers, since they are predictable and will fall exactly as listed above. These numbers are thereby called pseudo random numbers, where pseudo is another term for pretend.

Note, in the above, if the seed were changed to $w_0 = 30$, say, the sequence of random numbers would be the following:

25,0,3,18,29,20,7,6,1,8,11,26,5,28,15,14,9,16,19,2,13,4,23,22,17,24,27,10,21, 12,31,30

As the seed changes, another full cycle of 32 numbers is again attained.

The examples here are illustrated with m = 32, a small set of random values. But when m is large like $(2^{31} - 1)$, a very large sequence of random numbers is generated. In several simulation situations, it is useful to use the same sequence of random numbers, and therefore the same seed is appropriately applied. In other situations, the seed is changed on each run so that a different sequence of random numbers is used in the analysis.

Summary

The integrity of computer simulation models is only as good as the reliability of the random number generator that produces the stream of random numbers one after the other. The chapter describes the difficult task of developing an algorithm to generate random numbers that are statistically valid and have a large cycle length. The linear congruent method is currently the common way to generate the random numbers for a computer. The parameters of this method include the multiplier and the seed. Only a few multipliers are statistically recommended, and two popular ones in use for 32-bit word length computers are presented. Another parameter is the seed and this allows the analyst the choice of altering the sequence of random numbers with each run, and also when necessary, offers the choice of using the same sequence of random numbers from one run to another.

Chapter 3
Generating Random Variates

Introduction

Random variables are classified as discrete or continuous. Discrete is when the variable can take on a specified list of values, and continuous is when the variable can assume any value in a specified interval. The mathematical function that relates the values of the random variable with a probability is the probability distribution. When a value of the variable is randomly chosen according to the probability distribution, it is called a random variate. This chapter describes the common methods to generate random variates for random variables from various probability distributions. Two methods are in general use for this purpose, one is called the *Inverse Transform* method (IT), and the other is the *Accept-Reject* method (AR). The IT method is generally preferred assuming the distribution function transforms as needed. If the distribution is mathematically complicated and not easily transformed, the IT method becomes complicated and is not easily used. The AR method generally requires more steps than the IT method. The chapter presents various adaptations of these two methods.

For notation in this chapter, and in the entire book, when a continuous uniform variate falls equally in the range from zero to one, the notation will be u ~ U(0,1). In the examples of the book, when a variate of u ~ U(0,1) is obtained, for simplicity, only two or three decimals are used to show how the routine is run. Of course, in real simulation situations, the u variates with all decimals in place are needed.

Inverse Transform Method

Perhaps the most common way to generate a random variate for a random variable is by the inverse transform method. The method applies to continuous and discrete variables.

N.T. Thomopoulos, *Essentials of Monte Carlo Simulation: Statistical Methods for Building Simulation Models*, DOI 10.1007/978-1-4614-6022-0_3, © Springer Science+Business Media New York 2013

Continuous Variables

Suppose x is a continuous random variable with probability density f(x) for a $\leq x \leq$ b. The cumulative distribution function (cdf) of x becomes $F(x) = \int_a^x f(x)dx$ where $0 \leq F(x) \leq 1$. Since u $\sim$ U(0,1) and F(x) both range between 0 and 1, a random variate of u is generated and then F(x) is set to equal u, from which the associated value of x is found. The routine below describes the procedure:

1. Generate a standard uniform random variate u $\sim$ U(0,1).
2. Set F(x) = u.
3. Find the value of x that corresponds to F(x) = u, i.e., x = F^{-1}(u).
4. Return x.

The function F^{-1}(u) is called the *inverse function* of F(x) = u.

Example 3.1 Suppose x is a random variable with

$$f(x) = 0.125x \qquad \text{for } 0 \leq x \leq 4.$$

The associated cdf is below:

$$F(x) = 0.0625x^2 \qquad \text{for } 0 \leq x \leq 4.$$

To find a random variate of x, the inverse function of F(x) is derived as below:

1. Set u = F(x) = $0.0625x^2$.
2. Hence, x = F^{-1}(u) = $\sqrt{u/0.0625}$.
3. Generate a random variate, u $\sim$ U(0,1). Assume u = 0.71.
4. Compute x = $\sqrt{0.71/0.0625}$ = 3.370.
5. Return x = 3.370.

Discrete Variables

Now consider a discrete random variable, x_i, where i = 1, 2, ... with probability distribution $P(x_i)$ for i = 1, 2, ... The cumulative distribution function of x_i is $F(x_i) = P(x \leq x_i)$. To generate a random variate with the inverse transform method, the following routine is run:

1. Generate a random standard uniform variate, u $\sim$ U(0,1).
2. From $F(x_i)$, find the minimum i, say i_0, where u < $F(x_i)$.
3. Return x_{i_0}.

Example 3.2 Suppose a discrete random variable x with range, (0, 1, 2, 3) and probabilities, p(0) = 0.4, p(1) = 0.3, p(2) = 0.2 and p(3) = 0.1. Note the cumulative distribution function becomes: F(0) = 0.4, F(1) = 0.7, F(2) = 0.9 and F(3) = 1.0.

Assume a simulation model is in progress and a random variate of x is needed. To comply, the following routine is run:

1. Generate a random standard uniform, $u \sim U(0,1)$, say, $u = 0.548$.
2. Find the minimum x where $u < F(x)$. Note this is $x = 1$.
3. Return $x = 1$.

Accept-Reject Method

Consider x as a continuous random variable with density $f(x)$ for $a \leq x \leq b$. Further, for notation sake, let $\tilde{x}$ denote the mode of x, and thereby $f' = f(\tilde{x})$ is the density value at the mode of x. Note where all values of $f(x)$ will be equal or less than f'. To find a random variate of x, the following routine of five steps is run:

1. Generate two uniform random variates, $u \sim U(0,1)$, as u_1, u_2.
2. Find $x = a + u_1(b-a)$.
3. Compute $f(x)$.
4. If $u_2 < f(x)/f'$, then accept x and go to 5.
   Else, reject x and repeat steps 1–4.
5. Return x.

Example 3.3 Suppose x is a random variable with

$$f(x) = 0.125x \qquad \text{for } 0 \leq x \leq 4.$$

The mode is at $\tilde{x} = 4$ and thereby $f' = f(4) = 0.5$. So, now the four steps noted above are followed:

1. Generate $(u_1, u_2) = (0.26, 0{,}83)$, say.
2. $x = 0 + 0.26(4-0) = 1.04$.
3. $f(1.04) = 0.13$.
4. Since $0.83 > 0.13/0.50 = 0.26$, reject $x = 1.04$, and repeat the steps 1–4.

1. Generate $(u_1, u_2) = (0.72, 0{,}15)$, say.
2. $x = 0 + 0.72(4-0) = 2.88$.
3. $f(2.88) = 0.36$.
4. Since $0.15 < 0.36/0.50 = 0.72$, accept $x = 2.88$.
5. Return $x = 2.88$.

Truncated Variables

Sometimes, when the inverse transform method applies, the random variable of interest is a truncated portion of another random variable. For example, suppose x has density $f(x)$ for $a \leq x \leq b$, and $F(x)$ is the associated cumulative distribution function of x. But assume the variable of interest is only a portion of the original density where $c \leq x \leq d$ and the limits c, d are within the original limits of a and b, Therefore, $a \leq c \leq x \leq d \leq b$. Note the new density of this truncated variable becomes,

$$g(x) = f(x)/[F(d) - F(c)] \quad \text{for } c \leq x \leq d$$

To find a random variate of this truncated variable, the following routine is applied. Note, however, the routine listed below does not need the truncated density g(x) just described above:

1. Compute F(c) and F(d).
2. Generate a random uniform variate $u \sim U(0,1)$.
3. Find $v = F(c) + u[F(d) - F(c)]$.
   Note, $F(c) \leq v \leq F(d)$.
4. Set the cumulative distribution to v, i.e., $F(x) = v$.
5. Find the value of x that corresponds to $F(x) = v$, i.e., $x = F^{-1}(v)$.
6. Return x.

Example 3.4 Suppose x is a random variable with

$$f(x) = 0.125x \qquad \text{for } 0 \leq x \leq 4.$$

and recall the cdf below,

$$F(x) = 0.0625x^2 \qquad \text{for } 0 \leq x \leq 4.$$

Assume a random variate between the limits of 1 and 2 is required in a simulation analysis, i.e., $1 \leq x \leq 2$. To accomplish, the four steps below are followed:

1. Note, $F(1) = 0.0625$ and $F(2) = 0.2500$.
2. Generate a random $u \sim U(0,1)$, say $u = 0.63$.
3. $v = 0.0625 + 0.63[0.2500 - 0.0625] = 0.1806$.
4. $x = F^{-1}(0.1806) = \sqrt{0.1806/0.0625} = 1.70$.
5. Return $x = 1.70$.

Order Statistics

Suppose n samples are taken from a continuous random variable, x, with density f(x) and cumulative distribution F(x), and these are labeled as $(x_1, \ldots, x_n)$. Sorting the n samples from low to high yields, $x_{(1)} \leq x_{(2)} \leq \cdots \leq x_{(n)}$ where $x_{(i)}$ is the i-th lowest value in the sample.

Sorted Values

The notation y is here used to denote the i-th sorted value from the n samples of x. From order statistics, the probability density of y becomes:

$$g(y) = n!/[(i-1)!(n-i)!]f(y)F(y)^{i-1}[1 - F(y)]^{n-i}$$

Note, there is one value of $x = y$, (i-1) values of x smaller than y, and (n-i) values larger than y. See Rose and Smith, (2002) for more detail on order statistics.

Minimum Value

Suppose y is the smallest value of x, whereby, $y = \min(x_1, \ldots, x_n)$. The probability density of y becomes:

$$g(y) = nf(y) \left[1 - F(y)\right]^{n-1}$$

and the corresponding distribution function is:

$$G(y) = [1 - F(y)]^n$$

To generate a random variate for the minimum of n values from a continuous random variable x, the following routine is run:

1. Generate a random $u \sim U(0,1)$.
2. Set $G(y) = u$ and $F(y) = v$.
   Hence, $u = (1-v)^n$ and $v = [1-u^{1/n}]$.
3. Using the inverse transform method, compute $y = F^{-1}(v)$.
4. Return y.

Maximum Value

When w is the largest value of x, then $w = \max(x_1, \ldots, x_n)$. Hence, the probability density of w becomes:

$$g(w) = nf(w) \, F(w)^{n-1}$$

and the corresponding distribution function is,

$$G(w) = F(w)^n$$

To generate a random variate for the maximum of n values of a continuous random variable x, the following routine is run:

1. Generate a random $u \sim U(0,1)$.
2. Set $G(w) = u$ and $F(w) = v$.
   Hence, $u = v^n$.
   and $v = u^{1/n}$.
3. Using the inverse transform, compute $w = F^{-1}(v)$.
4. Return w.

Example 3.5 Suppose the density for a random variable x is $f(x) = 0.125x$ $(0 \leq x \leq 4)$, and thereby, the cumulative distribution is,

$$F(x) = 0.0625x^2 \qquad \text{for } 0 \leq x \leq 4.$$

Recall, also where the inverse function is

$$F^{-1}(u) = \sqrt{u/0.0625}$$

Suppose $n = 8$ samples of x are taken and of interest is to generate a random variate for the minimum value of the samples. To accomplish, the following steps are taken:

1. Generate a random variate $u \sim U(0,1)$, say $u = 0.37$.
2. $v = [1-0.37^{1/8}] = 0.117$.
3. $y = \sqrt{0.117/0.0625} = 1.367$.
4. Return $y = 1.367$.
   Note $F(1.367) = 0.117$.

Example 3.6 Suppose the density for a random variable x is $f(x) = 0.125x$ $(0 \leq x \leq 4)$, and thereby, the cumulative distribution is,

$$F(x) = 0.0625x^2 \qquad \text{for } 0 \leq x \leq 4.$$

Recall, also where the inverse function is

$$F^{-1}(u) = \sqrt{u/0.0625}$$

Suppose $n = 8$ samples of x are taken and of interest is to generate a random variate for the maximum value of the samples. To accomplish, the following steps are taken:

1. Generate a random variate $u \sim U(0,1)$, say $u = 0.28$.
2. $v = 0.28^{1/8} = 0.853$.
3. $w = \sqrt{0.853/0.0625} = 3.694$.
4. Return $w = 3.694$.
   Note $F(3.694) = 0.853$.

Composition

Sometimes the random variable is composed of a series of probability densities where each density occurs with a probability. This happens when there are k densities, $f_i(x)$, where the probability of density i being selected is p_i, and the sum of all the p_i is one. In essence, x is a random variable with probability density as below:

$$f(x) = p_1 f_1(x) + \ldots + p_k f_k(x)$$

$$\text{and } \sum_{i=1}^{k} p_i = 1$$

Note, where each of the k densities, $f_i(x)$, has a unique cumulative distribution function, $F_i(x)$, and a corresponding unique inverse function, $F_i^{-1}(u)$.

The composition can be described as below.

i	$f_i(x)$	p_i	G_i
1	$f_1(x)$	p_1	$G_1 = p_1$
...			
k	$f_k(x)$	p_k	$G_k = G_{k-1} + p_k$

The term G_i is the cumulative distribution function of the p_i's, and when $i = k$, $G_k = 1$.

Example 3.7 The density for variable x is composed of two separate densities, $f_1(x) = 1.25x$ for $(0 \leq x \leq 4)$ and $f_2(x) = 0.25$ for $(2 \leq x \leq 6)$. The associated probabilities are $p_1 = 0.6$ for $f_1(x)$, and $p_2 = .4$ for $f_2(x)$. So, $G_1 = 0.6$ and $G_2 = 1.0$. Note also, $F_1^{-1}(u) = \sqrt{u/0.0625}$, and $F_2^{-1}(u) = 2 + 4u$. To generate a random x, two random $u \sim U(0,1)$, are needed, u_1 and u_2. The steps below are followed:

1. Find two random uniform variates, $u \sim U(0,1)$. Say $(u_1, u_2) = (0.14, 0.53)$.
2. Since $u_1 < G_1 = 0.60$, density $f_1(x)$ is used.
3. $x = F_1^{-1}(0.53) = \sqrt{0.53/0.0625} = 2.91$.
4. Return x = 2.91.

Summation

On some occasions when generating a random variate, the sum of a random variable is needed, as in $y = (x_1 + \ldots + x_k)$. For notation convenience in this section, y will denote the sum of k independent samples of x, where x is a random variable with distribution $f(x)$. This method of summing is applied in subsequent chapters, in a convolution manner, to generate a random variate for the continuous Erlang distribution, and also for the discrete Pascal distribution.

To generate a random value of y from a continuous x with the inverse transform, $F^{-1}(u)$, known, the following loop is followed:

1. Set $y = 0$.
2. For $i = 1$ to k.
3. Generate a random $u_i \sim U(0,1)$.
4. $x_i = F^{-1}(u_i)$.
5. $y = y + x_i$.
6. Next i.
7. Return y.

Example 3.8 Suppose x is a continuous random variable with density $f(x) = 0.125x$ for $(0 \leq x \leq 4)$. The associated inverse function of x is $F^{-1}(u) = \sqrt{u/0.0625}$. Assume a sum of $k = 2$ is called and the result is another random variate y. The steps below show how one random observation of y is generated:

1. Generate $(u_1, u_2) = (0.44, 0.23)$, say.
2. $x_1 = F^{-1}(0.44) = \sqrt{0.44/0.0625} = 2.653$
   $y = 2.653$.
3. $x_2 = F^{-1}(0.23) = \sqrt{0.23/0.0625} = 1.918$.
   $y = 2.653 + 1.918 = 4.571$.
4. Return $y = 4.571$.

Triangular Distribution

The triangular is a continuous distribution that is sometimes used in simulation models when the true distribution of the variable is not known. To apply, the analyst estimates the minimum value of the variable, the maximum value, and the most likely value. These estimated values are denoted as: a, b and $\tilde{x}$, respectively. The random variable is labeled as x where $(a \leq x \leq b)$ and $(a \leq \tilde{x} \leq b)$.

Another variable is now introduced and follows the *standard triangular distribution*. The random variable is x' and this ranges from 0 to 1 where the most likely value (the mode) is labeled $\tilde{x}'$. So $(0 \leq x' \leq 1)$ and $(0 \leq \tilde{x}' \leq 1)$. The notation for the two variables, triangular x and standard triangular x', are below:

$$x \sim T(a, \ b, \tilde{x})$$

$$x' \sim T(0, \ 1, \tilde{x}')$$

The standard triangular x' is related to the triangular x as follows:

$$x' = (x - a)/(b - a)$$

and

$$\tilde{x}' = (\tilde{x} - a)/(b - a)$$

The probability density of x' is the following:

$$f(x') = 2x'/\tilde{x}' \qquad\qquad (0 \leq x' \leq \tilde{x}')$$
$$2(1 - x')/(1 - \tilde{x}') \qquad (\tilde{x}' < x' \leq 1)$$

and the associated cumulative distribution function becomes:

$$F(x') = x'^2/\tilde{x} \qquad\qquad (0 \leq x' \leq \tilde{x}')$$
$$1 - (1 - x')^2/(1 - \tilde{x}') \quad (\tilde{x}' < x' \leq 1)$$

The mean and variance of the standard triangular x' are below.

$$E(x') = (\tilde{x}' + 1)/3$$

$$V(x') = (1 + \tilde{x}'^2 - \tilde{x}')/18$$

The expected value and variance of the triangular x is related to the same from the standard triangular x' as shown below.

$$E(x) = a + E(x')[b - a]$$

$$V(x) = V(x')[b - a]^2$$

To find a random variate for a standard triangular x' ~ T(0,1,$\tilde{x}'$), the following routine is run:

1. Generate a random u ~ U(0,1).
2. If u $\leq \tilde{x}'$ x' $= \sqrt{u\tilde{x}'}$.
   If u $> \tilde{x}'$ x' $= 1 - \sqrt{(1 - \tilde{x}')(1 - u)}$.
3. Return x'.

Note the corresponding value of T(a,b,$\tilde{x}$) becomes x $= a + x'(b-a)$.

Example 3.9 Consider a variable x that is from a triangular distribution where x ranges from 20 to 60 and the most likely value is 30. When a random variate of x is needed in a simulation model, the following steps take place:

1. The mode $\tilde{x}'$ for the standard triangular distribution is computed by $\tilde{x}' = (30-20)/(60-20) = 0.25$.
2. A random uniform variate,u ~ U(0,1), is generated. Say u $= 0.38$.
3. Since u$>\tilde{x}'$, x' $= 1 - \sqrt{(1 - 0.25)(1 - 0.38)} = 0.318$.
4. x $= 20 + 0.318[60-20] = 32.72$.
5. Return x $= 32.72$.

Empirical Ungrouped Data

Sometimes, in simulation modeling, the data for a variable is not used to seek the theoretical continuous density, but instead is applied directly to define the distribution. The density that results is called the empirical distribution. Suppose the data is denoted as $(x_1, \ldots, x_n)$, and when sorted from low to high, it becomes $x_{(1)} \leq x_{(2)} \leq \ldots \leq x_{(n)}$. The data is then arranged in tabular form as below, along with the associated cumulative distribution function, $F_{x(i)}$.

i	x(i)	$F_{x(i)} = (i\text{-}1)/(n\text{-}1)$
1	x(1)	0
2	x(2)	1/(n-1)
...		
n	x(n)	1

To generate a random x, the composition method is used in the following way:

1. Generate two random uniform variates, $u \sim U(0,1)$, u_1 and u_2.
2. If $F_{x(i)} \leq u_1 < F_{x(i+1)}$, set $x = x(i) + u_2[x(i + 1){-}x(i)]$.
3. Return x.

Example 3.10 Suppose five observations of a continuous random variable are the following: 10, 5, 2, 20, 11. When sorted, they become: 2, 5, 10, 11, 20. The tabular form and the cumulative distribution function are listed below:

i	x(i)	$F_{x(i)}$
1	2	0.00
2	5	0.25
3	10	0.50
4	11	0.75
5	20	1.00

To generate a random x by the composition method, the following steps are followed:

1. Generate two random uniform variates, say $(u_1, u_2) = (0.82, 0.13)$.
2. Since $0.75 \leq 0.82 < 1.00$, $i = 4$, and $x = 11 + 0.13[20{-}11] = 12.17$.
3. Return $x = 12.17$.

Another way to generate the random variate is by the inverse transform method, as given below:

1. Generate a random uniform variate, $u \sim U(0,1)$.
2. If $F_{x(i)} \leq u < F_{x(i+1)}$, set $x = x(i) + \{[u{-}F_{x(i)}]/[F_{x(i+1)}{-}F_{x(i)}]\}[x(i + 1){-}x(i)]$.
3. Return x.

Example 3.11 Assume the same data (2, 5, 10, 11, 20) as Example 3.10.

To generate a random x by the inverse transform method, the following steps are followed:

1. Generate a random uniform variate, say u = 0.82.
2. Since $0.75 \leq 0.82 < 1.00$, i = 4, and
   $x = 11 + \{[0.82-0.75]/[1.00-0.75]\}[20-11] = 13.52$.
3. Return x = 13.52.

Empirical Grouped Data

Sometimes the data comes in grouped form as shown in the table below. Note the k intervals where $(a_i \leq x < b_i)$ identify the limits of x within each interval, and f_i is the frequency of samples in interval i. The sum of the frequencies is denoted as n. The cumulative distribution function for interval i becomes $F_i = F_{i-1} + f_i/n$, where $F_0 = 0$.

i	$[a_i, b_i)$	f_i	F_i
			$F_0 = 0$
1	$[a_1, b_1)$	f_1	$F_1 = F_0 + f_1/n$
2	$[a_2, b_2)$	f_2	$F_2 = F_1 + f_2/n$
...			
k	$[a_k, b_k)$	f_k	$F_k = F_{k-1} + f_k/n$

To generate a random x, the composition method is used in the following way:

1. Generate two random uniform variates, $u \sim U(0,1)$, u_1, u_2.
2. Find the interval, i, where $F_{i-1} \leq u_1 < F_i$, and set $x = a_i + u_2(b_i - a_i)$.
3. Return x.

Example 3.12 Suppose a variable to be used in a simulation study is presented in grouped form with the five intervals as listed below. Note, the table lists the range within each interval, the associated frequency and also the cumulative distribution. The sum of the frequencies is $n = 80$.

i	$[a_i, b_i)$	f_i	F_i
1	[5,8)	2	0.0250
2	[8,11)	27	0.3625
3	[11,13)	32	0.7625
4	[13,15)	15	0.9500
5	[15,18)	4	1.0000

To find a random x by the composition method, the routine below is applied:

1. Generate two random uniform variates, $u \sim U(0,1)$. Say $(u_1, u_2) = (0.91, 0.37)$.
2. Since $0.7625 \leq 0.91 < 0.9500$, i = 4 and $x = 13 + 0.37(15-13) = 13.74$.
3. Return x = 13.74.

Another way to generate the random variate is by the inverse transform method as shown below:

1. Generate a random uniform variate, $u \sim U(0,1)$.
2. Find the interval, i, where $F_{i-1} \leq u < F_i$, and set $x = a_i + \{[u-F_{i-1}]/[F_i-F_{i-1}]\}(b_i-a_i)$.
3. Return x.

Example 3.13 Assume the same data as in Example 3.12. To find a random x by the inverse transform method, the routine below applies:

1. Generate a random uniform variate, $u \sim U(0,1)$. Say $u = 0.91$.
2. Since $0.7625 \leq 0.91 < 0.9500$, $i = 4$ and
   $x = 13 + \{[0.91-0.7625]/[0.9500-0.7625]\}(15-13) = 14.573$.
3. Return $x = 14.573$.

Summary

This chapter shows how the continuous uniform $u \sim U(0,1)$ random variates are used to generate random variates for random variables from defined probability distributions. To accomplish in a computer simulation model, a random number generator algorithm is applied whenever a random uniform $u \sim U(0,1)$ variate is needed. The random number generator is the catalyst that delivers the uniform, $u \sim U(0,1)$, random variates, one after another, as they are needed in the simulation model. This is essential since it allows the analyst the opportunity to create simulation models that use any probability distribution that pertains and gives flexibility to emulate the actual system under study.

Chapter 4
Generating Continuous Random Variates

Introduction

A continuous random variable has a mathematical function that defines the relative likelihood that any value in a defined interval will occur by chance. The mathematical function is called the probability density. For example, the interval could be all values from 10 to 50, or might be all values zero or larger, and so forth. This chapter considers the more common continuous probability distributions and shows how to generate random variates for each. The probability distributions described here are the following: the continuous uniform, exponential, Erlang, gamma, beta, Weibull, normal, lognormal, chi-square, student's t, and Fishers F. Because the standard normal distribution is so useful in statistics and in simulation, and no closed-form formula is available, the chapter also lists the Hastings approximation formula that measures the relationship between the variable value and its associated cumulative probability.

Continuous Uniform

A variable x is defined as continuous uniform with parameters (a, b) when x is equally likely to fall anywhere from a to b. For example, suppose an experiment in a laboratory gives a temperature reading that rounds to $30°$ Fahrenheit. The true temperature, though, in three decimals, would be somewhere between $29.500°$ and $30.499°$. Hence, assume a continuous distribution where x has parameters a = 29.500 and b = 30.499.

The probability density of x is,

$$f(x) = 1/(b - a) \quad \text{for} \quad a \leq x \leq b$$

N.T. Thomopoulos, *Essentials of Monte Carlo Simulation: Statistical Methods for Building Simulation Models*, DOI 10.1007/978-1-4614-6022-0_4,
© Springer Science+Business Media New York 2013

and the cumulative distribution function becomes,

$$F(x) = (x - a)/(b - a) \quad \text{for } a \leq x \leq b$$

The expected value and the variance of x are the following:

$$E(x) = (b + a)/2$$

$$V(x) = (b - a)^2/12$$

To generate a random variate from the continuous uniform distribution, the following routine is run:

1. Generate a random uniform $u \sim U(0,1)$
2. $x = a + u(b - a)$
3. Return x

Example 4.1 Suppose x is continuous uniform with parameters (10, 20) and a random variate of x is needed. To accomplish, the following steps are followed:

1. Generate a random uniform $u \sim U(0,1)$, say $u = 0.68$
2. $x = 10 + 0.68(20 - 10) = 16.8$
3. Return x = 16.8

Exponential

The exponential distribution is heavily used in many applications and especially in queuing systems to define the time between units arriving to a system, and also for the times associated with servicing the units in the system. The probability density, f(x), is largest at x = 0 and continually decreases as x increases. The density has one parameter, θ, and is defined as below:

$$f(x) = \theta e^{-\theta x} \quad \text{for } x \geq 0$$

The associated cumulative distribution function becomes,

$$F(x) = 1 - e^{-\theta x} \quad \text{for } x \geq 0$$

The mean and variance of x are the following,

$$\mu = 1/\theta$$

and

$$\sigma^2 = 1/\theta^2$$

The inverse transform method is used to generate a random variate x in the following way. A random continuous uniform variate from u $\sim$ U(0,1) is obtained and is set equal to F(x) as below:

$$F(x) = u = 1 - e^{-\theta x}$$

Now, solving for x, yields the random variate of x by the relation,

$$x = -1/\theta \ln(1 - u)$$

where ln denotes the natural logarithm.

Standard Exponential

Note, the expected value of x is $E(x) = 1/\theta$, and in the special case when $\theta = 1$, the expected value of x is $E(x) = 1$. When $x = 1$ (the mean), $F(1) = 0.632$, indicating that 63.2 % of the values of x are below the mean value and 36.8 % are above. In this special situation at $E(x) = 1$, the distribution is like a standard exponential distribution. The list below relates some selective values of the cumulative distribution function, F(x), with the corresponding values of x.

F(x)	x
0.0	0.000
0.1	0.105
0.2	0.223
0.3	0.357
0.4	0.511
0.5	0.693
0.6	0.916
0.7	1.204
0.8	1.609
0.9	2.303

Note, the median occurs at x = 0.693 and the mean at x = 1.00, indicating the distribution skews far to the right.

Example 4.2 Assume an exponential distributed random variable x with a mean of 20 and a random variate of x is called in a simulation model. To generate the random x, the following steps are followed:

1. Generate a random uniform u $\sim$ U(0,1), say u = 0.17
2. $x = -20 \times \ln(1-0.17) = 3.727$
3. Return x = 3.727

Erlang

In some queuing systems, the time associated with arrivals and service times is assumed as an Erlang continuous random variable. The Erlang variable has two parameters, θ and k. The parameter θ is the scale parameter, and k, an integer, identifies the number of independent exponential variables that are summed together to form the Erlang variable. In this way, the Erlang variable x is related to the exponential variable y as below:

$$x = (y_1 + \ldots + y_k),$$

The expected value of x is related to the expected value of y as below:

$$E(x) = kE(y) = k/\theta$$

and the variance of x is derived from adding k variances of y, $V(y)$, as below:

$$V(x) = kV(y) = k/\theta^2$$

Note, when $k = 1$, the Erlang variable is the same as an exponential variable where the mode is zero and the density skews far to the right. As k increases, the shape of the Erlang density resembles a normal variable, via the central limit theorem.

The probability density of x is

$$f(x) = x^{k-1}\theta^k e^{-\theta x}/(k-1)! \qquad x \geq 0$$

and the cumulative distribution function is

$$F(x) = 1 - e^{-\theta x}\sum_{j=0}^{k-1}(\theta x)^j/j! \qquad x \geq 0$$

To generate a random Erlang variate of x with parameters, θ and k, the following routine is run:

1. Set $x = 0$
2. For $j = 1$ to k
3. Generate a random continuous uniform variate $u \sim U(0,1)$
4. $y = -(1/\theta)\ln(1 - u)$
5. Sum $x = x + y$
6. Next j
7. Return x

Example 4.3 Suppose a random Erlang variate is needed in a simulation run for variable x with parameter k = 4 and whose mean is 20. Note, because E(x) = 4/θ = 20, θ = 4/20 = 0.20. The following three steps yield the random x:

1. Generate four random continuous uniform u ~ U(0,1) variates, say (u_1, u_2, u_3, u_4) = (0.27, 0.69, 0.18, 0.76).
2. Using y = −(1/0.2)[ln(1 − u)], the corresponding random exponential variates are: (y_1, y_2, y_3, y_4) = (1.574, 5.856, 0.992, 7.136).
3. Summing the four exponentials yields: x = 14.658.
4. Return x = 14.658.

Gamma

The gamma distribution is almost the same as the Erlang distribution, except the parameter k is any value larger than zero whereas k is a positive integer for the Erlang. Also, x is any value greater or equal to zero. The density of the gamma is:

$$f(x) = x^{k-1}\theta^k e^{-\theta x}/\Gamma(k) \qquad x \geq 0$$

where $\Gamma(k)$ is called the gamma function, (not a density), defined as

$$\Gamma(k) = \int_0^\infty t^{k-1}e^{-t}dt \qquad \text{for } k>0$$

When k is a positive integer, $\Gamma(k) = (k − 1)!$
The mean and variance of x are the following:

$$\mu = k/\theta$$

and

$$\sigma^2 = k/\theta^2$$

To generate a random gamma variate is not easy. The method presented here depends on whether k < 1 or k > 1. Note, when, k = 1, the distribution is the same as an exponential.

When k < 1

A routine to generate a random x, when k < 1, comes from Ahrens and Dieter (1974). It is based on the Accept-Reject method and is shown below in five steps. In the computations below, x' is a gamma variate with θ = 1, and x is a gamma variate with any positive θ:

1. Set b $= (e + k)/e$ where e ≈ 2.71828 is the base of the natural logarithm.
2. Generate two random uniform u $\sim U(0,1)$variates, u_1 and u_2.
   $p = bu_1$
   if $p > 1$ go to step 4
   if $p \leq 1$ go to step 3
3. $y = p^{1/k}$
   if $u_2 \leq e^{-y}$, set $x' = y$, go to step 5
   if $u_2 > e^{-y}$, go to step 2
4. $y = -\ln[(b - p)/k]$
   if $u_2 \leq y^{k-1}$, set $x' = y$, go to step 5
   if $u_2 > y^{k-1}$, go to step 2
5. Return $x = x'/\theta$.

When $k > 1$

Cheng (1977) developed the routine to generate a random x when $k > 1$. The method uses the Accept-Reject method as shown in the five steps listed below. Note below where x' is a gamma variate with $\theta = 1$, and x is a gamma variate with any positive θ:

1. Set a $= 1/\sqrt{2k - 1}$
   $b = k - \ln4$, where $\ln$ = natural logarithm.
   $q = k + 1/a$
   $c = 4.5$
   $d = 1 + \ln(4.5)$
2. Generate two random uniform u $\sim U(0,1)$ varites, u_1 and u_2.
   $v = a \times \ln[u_1/(1 - u_1)]$
   $y = ke^v$
   $z = u_1^2 u_2$
   $w = b + qv - y$
3. if $w + d - cz \geq 0$, set $x' = y$, go to step 5
   if $w + d - cz < 0$, go to step 4
4. if $w \geq \ln(z)$, set $x' = y$, go to step 5
   If $w < \ln(z)$, goto step 2
5. Return $x = x'/\theta$.

Example 4.4 Suppose x is gamma distributed with a mean of 0.1 and the variance $= 0.02$. Since, $\mu = k/\theta$, and $\sigma^2 = k/\theta^2$, then solving for k and θ, yields $k = 0.5$ and $\theta = 5$. The computations are below:

1. e $= 2.718$
   $b = 1.184$
2. (u1, u2) $= (0.71, 0.21)$, say.
   Since $p = 0.841 \leq 1$, go to step 3

3. $y = 0.707$
   $e^{-y} = 0.493$
   Since $u2 \leq 0.493$, $x' = 0.707$
4. Return $x = 0.707/5 = 0.141$.

Example 4.5 Suppose x is gamma distributed with a mean of 10 and the variance $=$ 66.6. Since, $\mu = k/\theta$, and $\sigma^2 = k/\theta^2$, then solving for k and θ, yields $k = 1.5$ and $\theta \approx 0.15$. The computations are below:

1. $a = 0.707$
   $b = 0.114$
   $q = 2.914$
   $c = 4.5$
   $d = 2.504$
2. $(u1, u2) = (0.15, 0.74)$, say
   $v = -1.226$
   $y = 0.440$
   $z = 0.0167$
   $w = -3.899$
3. Since $w + d - cz < 0$, go to 4
4. $\ln(z) = -4.092$
   Since $w \geq \ln(z)$, $x' = 0.440$
5. Return $x = 0.440/0.15 = 2.933$

Beta

The beta distribution has two parameters (k_1, k_2) where $k_1 > 0$ and $k_2 > 0$, and takes on many shapes depending on the values of the parameters. The variable denoted as x, lies within two limits, a and b where $(a \leq x \leq b)$.

Standard Beta

Another distribution is introduced and is called the standard beta. This distribution has the same parameters (k_1, k_2) as the beta distribution, but the limits are constrained to the range $(0,1)$. So when x is a beta with a range (a,b), x' is a standard beta with a range $(0,1)$. When they both have the same parameters, x and x' are related as below:

$$x' = (x - a)/(b - a)$$

and

$$x = a + x'(b - a)$$

The probability density for x' is the following:

$$f(x') = (x')^{k_1-1}(1 - x')^{k_2-1}/B(k_1, k_2) \qquad\qquad (0 \le x' \le 1)$$

where

$$B(c, d) = \text{beta function} = \int_0^1 t^{c-1}(1 - t)^{d-1} dt$$

The expected value of x' is

$$E(x') = k_1/(k_1 + k_2)$$

and the variance is

$$V(x') = k_1 k_2 \Big/ \left[(k_1 + k_2)^2(k_1 + k_2 + 1) \right]$$

The corresponding expected value and variance of x becomes,

$$E(x) = a + E(x')(b - a)$$

$$V(x) = (b - a)^2 V(x')$$

The mode of the standard beta variable could be 0 or 1 depending on the values of k_1 and k_2. However, when $k_1 > 1$ and $k_2 > 1$, the mode lies between 0 and 1 and is computed by,

$$\tilde{x}' = (k_1 - 1)/(k_1 + k_2 - 2)$$

The mode for the beta variable becomes:

$$\tilde{x} = a + \tilde{x}'(b - a)$$

Below is a list describing the relation between the parameters and the shape of the distribution.

Parameters	Shape
$k_1 < 1$ and $k_2 \ge 1$	Mode at zero (right skewed)
$k_1 \ge 1$ and $k_2 < 1$	Mode at one (left skewed)
$k_1 = 1$ and $k_2 > 1$	Ramp down from zero to one
$k_1 > 1$ and $k_2 = 1$	Ramp up from zero to one
$k_1 < 1$ and $k_2 < 1$	Bathtub shape
$k_1 > 1$ and $k_2 > 1$ and $k_1 > k_2$	Mode between zero and one (left skewed)
$k_1 > 1$ and $k_2 > 1$ and $k_2 > k_1$	Mode between zero and one (right skewed)
$k_1 > 1$ and $k_2 > 1$ and $k_1 = k_2$	Mode in middle, symmetrical, normal like
$k_1 = k_2 = 1$	Uniform

To generate a random beta variate x with parameters (k_1, k_2), and with the range (a, b) the following routine is run:

1. Generate a random gamma variate, g_1, with parameters k_1 and $\theta_1 = 1$.
   Generate a random gamma variate, g_2, with parameters k_2 and $\theta_2 = 1$.
2. $x' = g_1/(g_1 + g_2)$
3. $x = a + x'(b - a)$
4. Return x

Example 4.6 Suppose x is a beta random variable with parameters (2,4) and has a range of 10–50. The following steps are followed to show how to generate a random x:

1. A random gamma variate is generated with $(k_1 = 2, \theta_1 = 1)$, say, $g_1 = 1.71$.
   A random gamma variate is generated with $(k_2 = 4, \theta_2 = 1)$, say, $g_2 = 4.01$.
2. The random standard beta variate is $x' = 1.71/(1.71 + 4.01) = 0.299$
3. The random beta variate is $x = 10 + 0.299(50 - 10) = 21.958$
4. Return $x = 21.958$

Weibull

The Weibull distribution has two parameters, $k_1 > 0$ and $k_2 > 0$, and the random variable, denoted as x, ranges from zero and above. The density is

$$f(x) = k_1 k_2^{-k_1} x^{k_1} \exp\left[-(x/k_2)^{k_1}\right] \qquad x > 0$$

and the cumulative distribution function

$$F(x) = 1 - \exp\left[-(x/k_2)^{k_1}\right] \qquad x > 0$$

The expected value and variance of x are listed below,

$$E(x) = k_2/k_1 \Gamma(1/k_1)$$

$$V(x) = k_2^2/k_1 \left[2\Gamma(2/k_1) - 1/k_1\Gamma(1/k_1)^2\right]$$

Recall Γ denotes the gamma function described earlier in this chapter. When the parameter $k_1 \leq 1$, the shape of the density is exponential like. When $k_1 > 1$, the shape has a mode greater than zero and skews to the right, and at $k_1 \geq 3$, the density shape starts looking like a normal distribution.

To generate a random x from the Weibull, the inverse transform method is used. Setting a random uniform variate $u \sim U(0,1)$ to F(x), and solving for x, yields the following:

$$x = k_2[-\ln(1 - u)]^{1/k_1}$$

Example 4.7 Suppose x is Weibull distributed with parameters $k_1 = 4$, $k_2 = 10$, and a random x is called in a simulation run. The following steps are followed:

1. Generate a random uniform $u \sim U(0,1)$. Say $u = 0.92$.
2. $x = 10[-\ln(1 - 0.92)]^{1/4} = 12.61$.
3. Return $x = 12.61$.

Normal Distribution

The normal distribution is symmetrical with a bell shaped density. Its mean is denoted as μ and the standard deviation as σ. This is perhaps the most widely used probability distribution in business and scientific applications. A companion distribution, the standard normal distribution, has a mean of zero, a standard deviation of one, and has the same shape as the normal distribution. The notation for the normal variable is $x \sim N(\mu,\sigma^2)$, and its counterpart, the standard normal is $z \sim N(0,1)$. The variable z is related to x by the relation: $z = (x-\mu)/\sigma$. In the same way, x is obtained from z by: $x = \mu + z\sigma$. When k represents a particular value of z, the probability density is $f(k) = 1/\sqrt{2\pi}\exp(-k^2/2)$. The probability that z is less than k is denoted as F(k) and is computed by $F(k) = \int_{-\infty}^{k} f(z)dz$.

There is no closed-form solution for the cumulative distribution F(z). A way to approximate F(z) has been developed by C.Hastings, Jr. (1955), and also reported by A. Abramowitz and I. A. Stegun (1964). Two methods credited to Hastings are listed below.

Hastings Approximation of F(z) from z

To find F(z) from a particular value of z, the following routine is run:

1. $d_1 = 0.0498673470$
   $d_2 = 0.0211410061$
   $d_3 = 0.0032776263$
   $d_4 = 0.0000380036$
   $d_5 = 0.0000488906$
   $d_6 = 0.0000053830$
2. If $z \geq 0$, $k = z$
   If $z < 0$, $k = -z$
3. $F = 1 - 0.5[1 + d_1k + d_2k^2 + d_3k^3 + d_4k^4 + d_5k^5 + d_6k^6]^{-16}$
4. if $z \geq 0$, $F(z) = F$

If z < 0, F(z) = 1 − F
5. Return F(z)

Hastings Approximation of z from F(z)

Another useful approximation also comes from Hastings, and gives a formula that yields a random z from a value of F(z). The routine is listed below:

1. $c_0 = 2.515517$
   $c_1 = 0.802853$
   $c_2 = 0.010328$
   $d_1 = 1.432788$
   $d_2 = 0.189269$
   $d_3 = 0.001308$
2. $H(z) = 1 − F(z)$
   If $H(z) \leq 0.5$, $H = H(z)$
   If $H(z) > 0.5$, $H = 1 − H(z)$
3. $t = \sqrt{\ln(1/H^2)}$ where ln = natural logarithm
4. $k = t − [c_0 + c_1 t + c_2 t^2]/[1 + d_1 t + d_2 t^2 + d_3 t^3]$
5. If $H(z) \leq 0.5$, $z = k$
   If $H(z) > 0.5$, $z = −k$
6. Return z

The literature reports various ways to generate a random standard normal variate z. Three of the methods are presented here.

Hastings Method

The first way utilizes the Hastings method that finds a z from F(z), and is based on the inverse transform method. The routine uses one standard uniform variate, $u \sim U(0,1)$, as shown below:

1. Generate a random continuous uniform variate $u \sim U(0,1)$.
2. Set F(z) = u.
3. Use Hastings Approximation of z from F(z) to generate a random standard normal variate z.
4. Return z.

Convolution Method

A second way to generate a random standard normal variate uses twelve random continuous uniform variates. The routine is listed below:

1. $z = -6$
2. For $i = 1$ to 12
3. Generate a random uniform variate $u \sim U(0,1)$
4. $z = z + u$
5. Next i
6. Return z

Note in the above routine, $E(u) = 0.5$ and $V(u) = 1/12$, and thereby, $E(z) = 0$ and $V(z) = 1$. Also since z is based on the convolution of twelve continuous uniform $u \sim U(0,1)$ variates, the central limit theorem applies and hence, $z \sim N(0,1)$.

Example 4.7 The routine below shows how to use the convolution method to generate a random $z \sim N(0.1)$:

1. Set $z = -6.0$
2. Sum 12 random continuous uniform variates, $u \sim U(0,1)$, say $\Sigma u = 7.12$
3. $z = -6.0 + \Sigma u = 1.12$
4. Return $z = 1.12$

Sine-Cosine Method

A third way generates two random standard normal variates, z_1, z_2. This method comes from a paper by Box and Muller (1958). The routine requires two random continuous uniform variates to generate the two random standard normal variates. The routine is listed below:

1. Generate two random continuous uniform variates, u_1 and u_2
2. $z_1 = \sqrt{-2\ln(u_1)} \cos[2\pi(u_2)]$

   $z_2 = \sqrt{-2\ln(u_1)} \sin[2\pi(u_2)]$
3. Return z_1 and z_2

Example 4.8 Suppose x is normally distributed with mean 40 and standard deviation 10, and a random normal variate of x is needed. Using the Sine-Cosine method, the steps below follow:

1. Two random continuous uniform $u \sim U(0,1)$ variates are $(u_1, u_2) = (.37, .54)$, say
2. $z_1 = \sqrt{-2\ln(u_1)} \cos[2\pi(u_2)] = -1.3658$

   $z_2 = \sqrt{-2\ln(u_1)} \sin[2\pi(u_2)] = -0.3506$
3. $x = 40 - 1.3658 \times 10 = 26.342$
4. Return $x = 26.342$.

Lognormal

The lognormal distribution with variable x > 0, reaches a peak greater than zero and skews far to the right. This variable is related to a counterpart normal variable y, in the following way.

$$y = \ln(x)$$

where ln is the natural logarithm. In the same way, x is related to y by the relation below:

$$x = e^y.$$

The variable y is normally distributed with mean and variance, μ_y and $\sigma_y{}^2$, respectively, and x is lognormal with mean and variance, μ_x and $\sigma_x{}^2$. The notation for x and y are as below:

$$x \sim LN(\mu_y, \sigma_y{}^2)$$

$$y \sim N(\mu_y, \sigma_y{}^2)$$

Note, the parameters to define the distribution of x, are the mean and variance of y. The parameters between x and y are related in the following way:

$$\mu_x = \exp\left[\mu_y + \sigma_y{}^2/2\right]$$

$$\sigma_x{}^2 = \exp\left[2\mu_y + \sigma_y{}^2\right]\left[\exp(\sigma_y{}^2) - 1\right]$$

$$\mu_y = \ln\left[\mu_x{}^2/\sqrt{\mu_x^2 + \sigma_x^2}\right]$$

$$\sigma_y{}^2 = \ln\left[1 + \sigma_x{}^2/\mu_x{}^2\right]$$

To generate a random x with parameters μ_y and $\sigma_y{}^2$, the following routine is run:

1. Generate a random standard normal variate, z.
2. A random normal variate becomes: $y = \mu_y + z\sigma_y$.
3. The random lognormal variate is $x = e^y$.
4. Return x.

Example 4.9 Suppose x is lognormal with mean $\mu_x = 10$, variance $\sigma_x{}^2 = 400$, and a random lognormal variate of x is needed. The steps below show how to find a random x:

1. The mean and variance of y become:

$$\mu_y = \ln\left[\mu_x{}^2/\sqrt{\mu_x^2 + \sigma_x^2}\right] = 1.498$$

$$\sigma_y{}^2 = \ln[1 + \sigma_x{}^2/\mu_x{}^2] = 1.609$$

The standard deviation of y is $\sigma_y = 1.269$.

2. A random standard normal variate is generated as $z = 1.37$, say.
3. The random normal variate becomes $y = 1.498 + 1.37 \times 1.269 = 3.236$.
4. The random lognormal variate is $x = e^{3.236} = 25.44$.
5. Return $x = 25.44$.

Chi-Square

The chi-square distribution is one of the most frequently used distributions in statistical analysis, usually to test the variability of the variance of a variable. The chi-square variable is denoted as χ^2 and is associated with a parameter k, the degrees of freedom. The variable χ^2 is greater or equal to zero, and the parameter k is a positive integer. When, $w = \chi^2$, the probability density of w, with parameter k, is listed below:

$$f(w) = [w^{(k/2-1)}e^{-w/2}]/\left[2^{k/2}\Gamma(k/2)\right] \quad w \geq 0$$

The mean and variance of w (and χ^2) are $E(w) = k$ and $V(w) = 2k$, respectively. So, the mean and variance of χ^2 with k degrees of freedom are:

$$E(\chi^2) = k$$

$$V(\chi^2) = 2k$$

Probability tables values of chi-square with parameter k and probability $P(\chi^2 > \chi^2_\alpha) = \alpha$ are listed in most statistical books, and usually for $k \leq 100$.

The variable chi-square with degrees of freedom k is related to the standard normal variable as shown below:

$$\chi^2 = z_1{}^2 + \ldots + z_k{}^2$$

where z_1 to z_k are standard normal variates.

Approximation Formula

When the parameter k is large (k $>$30), thanks to the central limit theorem, the chi-square probability density is shaped like a normal distribution whereby, $\chi^2 \sim$ N(k,2k). Using this relation, an approximation to the α-percent chi-square value is shown below:

$$\chi^2_\alpha \approx k + z_\alpha \sqrt{2k}$$

where z is a standard normal variable with P(z $>$ z_α) $= \alpha$ and thereby $P(\chi^2 > \chi^2_\alpha) \approx \alpha$.

Relation to Gamma

It is also noted where the density f(w) has the same shape as the gamma distribution with parameters θ and k$'$ when the gamma parameters are set as: $\theta = 2$ and k$' = $k/2,

Generate a Random Chi-Square Variate

A random chi-square variate with degrees of freedom, k, can be generated by summing k standard normal variates as shown in the routine below:

1. $\chi^2 = 0$
2. For i $= 1$ to k
3. Generate a standard normal variate z
4. $\chi^2 = \chi^2 + z^2$
5. Next i
6. Return χ^2

Another way to generate the chi-square variate is by using the gamma relationship noted above. For chi-square with parameter k, generate a gamma with parameters, (2, k/2) and the outcome becomes the random chi-square variate.

When k is large, (k $>$ 30), the normal approximation given above, can be used to generate the chi-square variate. In this situation the chi-square variate is approximated by the normal distribution with a mean k and variance 2k.

Example 4.10 Suppose a random chi-square with degrees of freedom k $= 3$ is needed in a simulation run. To generate a random chi-square, the steps below are followed:

1. Suppose three random standard normal variates are: $(z_1, z_2, z_3) = (0.47, -0.81, 1.04)$, say.
2. $\chi^2 = 0.47^2 + -0.81^2 + 1.04^2 = 1.9586$.
3. Return $\chi^2 = 1.9586$.

Example 4.11 Suppose a simulation model needs a chi-square random variate with 239 degrees of freedom. To accomplish, the following routine is run:

1. Generate a random standard normal $z \sim N(0,1)$, say $z = 1.34$.
2. $\chi^2 = \text{int}\left(239 + 1.34\sqrt{2 \times 239} + 0.5\right) = 269$
3. Return $\chi^2 = 269$

Student's t

The student's t distribution is an important distribution used in statistical analysis, usually to test the significance of the mean value of a variable. The distribution is often referred as the t distribution. The spread of the distribution is much like the standard normal distribution but the tails can reach out farther to the right and left, depending on a parameter k, the degrees of freedom. The expected value and variance of t are listed below:

$$E(t) = 0$$

$$V(t) = k/(k - 2) \quad \text{at } k>2$$

When $k > 30$, the student's t distribution is approximated by the standard normal distribution.

The variable t, with parameter k, is related to the standard normal distribution and the chi-square distribution by the relation below:

$$t = z/\sqrt{\chi_k^2/k}$$

Generate a Random Variate

To generate a random t with parameter k, the following routine is run:

1. Generate a random standard normal variate, $z \sim N(0,1)$.
2. Generate a random chi-square variate with parameter k, χ_k^2.
3. $t = z/\sqrt{\chi_k^2/k}$
4. Return t.

Example 4.12 Suppose a random variate t with degrees of freedom $k = 6$ is needed in a simulation analysis. To accomplish, the steps below are followed.

1. A random standard normal variate is generated as $z = 0.71$, say.
2. A random chi-square variate with parameter $k = 6$ is generated as $\chi_6^2 = 6.29$, say.
3. $t = 0.71/\sqrt{6.29/6} = 0.693$
4. Return $t = 0.693$.

Fishers' F

Fisher's F distribution is an important distribution used in statistical analysis and pertains when the two or more variances from normal variables are under review. The variable F is greater than zero, and has two parameters, u and v, where both are positive integers. The variable F is derived from two independent chi square variables, χ_1^2 and χ_2^2 with degree of freedom, u and v, respectively, as shown below.

$$F = \left[\chi_1^2/u\right]/\left[\chi_2^2/v\right] \text{ and } F>0$$

The expected value and variance of F are listed below,

$$E(F) = v/(v-2) \qquad\qquad\qquad\qquad \text{when } v>2$$

$$V(F) = \left[2v^2(u+v-2)\right]/\left[u(v-2)^2(v-4)\right] \qquad \text{when } v>4$$

Table values, $F\alpha_{,u,v}$, of F with parameters u and v are listed in most statistical books where $P[F > F\alpha_{,u,}] = \alpha$. Note the relation below that shows how the lower tail values of F with degrees of freedom v and u is related to the upper tail values when the degrees of freedom are u and v.

$$F_{(1-\alpha),v,u} = 1/F_{\alpha,u,v}$$

In statistical analysis, suppose x_1 and x_2 are two normally distributed variables with variances σ_1^2 and σ_2^2, respectively, and s_1^2 and s_2^2 are the corresponding sample variances when n_1 and n_2 are the number of samples taken for x_1 and x_2, respectively. The ratio,

$$F = \left[s_1^2/\sigma_1^2\right]/\left[s_2^2/\sigma_2^2\right]$$

is distributed as an F variable with degrees of freedom $n_1' = n_1-1$ and $n_2' = n_2-1$. When σ_1^2 and σ_2^2 are equal, the ratio becomes,

$$F = s_1^2/s_2^2$$

Note, $u = n_1'$ and $v = n_2'$ is the notation for the degrees of freedom. To generate a random F with parameters, u and v, the following routine is run:

1. Generate χ_1^2 and χ_2^2 with degree of freedom, u and v, respectively.
2. $F = [\chi_1^2/u]/[\chi_2^2/v]$
3. Return F

Example 4.13 Suppose a random variate of F with degrees of freedom 4 and 6 is needed in a simulation run. The steps below show how the random F is derived:

1. Suppose $\chi_1^2 = 3.4$ and $\chi_2^2 = 8.3$, are randomly generated.
2. $F = [3.4/4]/[8.3/6] = 0.61$
3. Return $F = 0.61$

Summary

This chapter shows how to transform the continuous uniform random variates, $u \sim U(0,1)$, to random variates for a variable that comes from one of the common continuous probability distributions. The probability distributions described here are the following: the continuous uniform, exponential, Erlang, gamma, beta, Weibull, normal, lognormal, chi-square, student's t, and Fishers F. The chapter also shows how to use the (Hastings) approximation formulas for the standard normal distribution.

Chapter 5
Generating Discrete Random Variates

Introduction

A discrete random variable includes a specified list of exact values where each is assigned a probability of occurring by chance. The variable can take on a particular set of discrete events, like tossing a coin (head or tail), or rolling a die (1,2,3,4,5,6). This chapter considers the more common discrete probability distributions and shows how to generate random variates for each. The probability distributions described here are the following: discrete arbitrary, discrete uniform, Bernoulli, binomial, hyper geometric, geometric, Pascal and Poisson.

Discrete Arbitrary

A variable x is defined as discrete when a set number of values of x can occur, as x_i for $i = 1$ to N, and N could be finite or infinite. Generally, x_i are the positive integers as $x_i = 0, 1, 2,\ldots$. The probability of a particular value x_i is denoted as $P(x_i) = P(x = x_i)$. Hence, $P(x_1), \ldots, P(x_N)$ define the probability distribution of x. The sum of all the probabilities is equal to one, i.e.,

$$\sum_i P(x_i) = 1$$

The expected value of x is obtained as below:

$$E(x) = \sum_i x_i P(x_i)$$

and the associated variance is,

$$V(x) = E(x^2) - E(x)^2$$

N.T. Thomopoulos, *Essentials of Monte Carlo Simulation: Statistical Methods for Building Simulation Models*, DOI 10.1007/978-1-4614-6022-0_5,
© Springer Science+Business Media New York 2013

where,

$$E(x^2) = \sum_i x_i^2 P(x_i)$$

The cumulative distribution function of x is denoted as $F(x_i)$ and is computed by

$$F(x_i) = P(x \le x_i)$$

To generate a random variate x from an arbitrary probability distribution, the following routine is run:

1. For each x_i, find $F(x_i)$ i = 1 to N.
2. Generate a random continuous uniform u ~ U(0,1).
3. Locate the smallest x_i where u < $F(x_i)$.
4. Set x = x_j.
5. Return x.

Example 5.1   Suppose x is discrete with the following probability distribution, and the associated cumulative distribution function.

x	P(x)	F(x)
0	0.4	0.4
1	0.3	0.7
2	0.2	0.9
3	0.1	1.0

To obtain a random x, the following steps are taken:

1. Generate a random u ~ U(0,1), say u = 0.57.
2. Note, x = 1 is the smallest x where u = 0.57 < F(1) = 0.7.
3. Return x = 1.

Discrete Uniform

A variable x follows the discrete uniform distribution with parameter (a,b) when x takes on all integers from a to b with equal probabilities. The probability of x becomes,

$$p(x) = 1/(b - a + 1) \qquad x = a \text{ to } b$$

The cumulative distribution function is

$$F(x) = (x - a + 1)/(b - a + 1) \qquad x = a \text{ to } b$$

The expected value and the variance of x are listed below:

$$E(x) = (a + b)/2$$

$$V(x) = \left[(b - a + 1)^2 - 1\right]/12$$

To generate a random discrete uniform variate of x, the routine below is followed:

1. Generate a random continuous uniform u $\sim$ U(0,1).
2. x = ceiling [(a−1) + u(b−a + 1)].
3. Return x.

Example 5.2 Suppose an analyst needs a random discrete uniform variate for use in a simulation model in progress where x includes all integers from 10 to 20. To compute, the following steps are taken:

1. Generate a continuous uniform random u $\sim$ U(0,1), say u = 0.714.
2. Calculate x = ceiling [(10−1) + 0.714 (20−10 + 1)] = ceiling [16.854] = 17.
3. Return x = 17.

Bernoulli

Suppose the variable x is distributed as a Bernoulli variable of x = 0 or 1, where the probability of each is the following:

$$P(x = 0) = 1 - p$$

$$P(x = 1) = p$$

The expected value and variance of x are the following:

$$E(x) = p$$

$$V(x) = p(1 - p)$$

To generate a random Bernoulli variate x, the following three steps are taken:

1. Generate a random uniform u $\sim$ U(0,1).
2. If u < p, x = 1; else, x = 0.
3. Return x.

Example 5.3 Consider a Bernoulli x with p = 0.70. A random x is generated as follows:

1. Generate a random u ~ U(0,1), say u = 0.48.
2. Since u < p = 0.70, x = 1.
3. Return x = 1.

Binomial

The variable x is distributed as a Binomial when x is the number of success' in n independent trials of an experiment with p the probability of a success per trial. The variable x can take on the integer values of 0–n. The probability of x is the following:

$$P(x) = n!/[x!(n - x)!]p^x(1 - p)^{n-x} \quad x = 0, \ldots, n$$

The expected value and variance of x are listed below:

$$E(x) = np$$

$$V(x) = np(1 - p)$$

The cumulative distribution function of x, denoted as F(x), is the probability of the variable achieving the value of x or smaller. When x = x_o, say,

$$F(x_o) = P(x \leq x_o).$$

To generate a random binomial variate of x, one of the three routines listed below may be used.

When n is Small

When n is small to moderate in size, the following routine is efficient:

1. Set x = 0.
2. For i = 1 to n
   Generate a random continuous uniform variate, u ~ U(0,1)
   If u < p, x = x + 1
   Next i.
3. Return x.

Normal Approximation

When n is large and if $p \leq 0.5$ with $np > 5$, or if $p > 0.5$ with $n(1-p) > 5$, then x can be approximated with the normal distribution, whereby $x \sim N[np, np(1-p)]$. The routine listed below will generate a random x:

1. Generate a random standard normal variate, $z \sim N(0,1)$.
2. $x = \text{integer}\left[np + z\sqrt{np(1-p)} + 0.5\right]$.
3. Return x.

Poisson Approximation

When n is large and p is small, and the above normal approximation does not apply, x is approximated by the Poisson distribution that is described subsequently in this chapter:

1. The expected value of the Poisson variable is denoted here as θ, where $E(x) = \theta = np$.
2. Generate a random Poisson variate x with parameter θ.
3. Return x.

Example 5.4 Suppose x is binomial distributed with $n = 5$ trials and $p = 0.375$. To generate a random binomial x, five continuous uniform variates, $u \sim U(0,1)$, are needed as shown below:

1. Suppose the five uniform variates are the following: 0.286, 0.949, 0.710, 0.633, and 0.325.
2. Since two of the variates are below $p = 0.375$, $x = 2$.
3. Return $x = 2$.

Example 5.5 Assume 100 independent Bernoulli trials are run where the probability of a success per trial is $p = 0.40$. Of interest is to generate a random binomial variate x. Since $n = 100$, $p = 0.40$ and $np = 40$, the normal approximation to the binomial can be used. The mean and variance of the normal are, $\mu = np = 40$ and $\sigma^2 = np(1-p) = 24$, respectively. Hence $\sigma = 4.89$. The following routine generates the random x:

1. Generate a random standard normal variate z, say $z = 0.87$.
2. $x = \text{integer}[\mu + z\sigma + 0.5] = 44$.
3. Return $x = 44$.

Example 5.6 Suppose a random binomial variate x is needed from a sample of $n = 1,000$ trials with $p = 0.001$. With $np = 1.0$, the normal distribution does not apply, but the Poisson distribution is applicable with $\theta = 1.00$. Later in this chapter, the way to generate a random x from the Poisson distribution is shown:

1. Generate a random Poisson variate with parameter $\theta = 1.00$, say $x = 2$.
2. Return $x = 2$.

Hyper Geometric

The variable x is distributed as a hyper geometric when x is the number of defectives in n samples taken without replacement from a population of size N with D defectives. The variable x can take on the integer values of zero to the smaller of D and n. The probability of x is the following:

$$P(x) = \binom{N-D}{n-x}\binom{D}{x} \bigg/ \binom{N}{n} \, x = 0, \ldots, \, \min(n, D)$$

The expected value and variance of x are listed below:

$$E(x) = nD/N$$

$$V(x) = n[D/N][1 - D/N][N - n]/[N - 1]$$

To generate a random hyper geometric variate, the following routine is run. The parameter notations are N = population size, D = number of defects in the population, and n = number of samples without replacement:

1. Set N1 = N, D1 = D and x = 0.
2. For i = 1 to n
   p = D1/N1
   Generate a random continuous uniform u ~ U(0,1)
   N1 = N1 - 1
   If u < p, x = x + 1 and D1 = D1−1
   Next i.
3. Return x.

Example 5.7 Suppose a situation where a lot of ten units contains two defectives and a sample of size four is taken without replacement. The goal is to generate a random hyper geometric variate on the number of defective units, x, observed in the sample. Note, N = 10, D = 2 and n = 4. The steps below show how a random x is generated:

1. Set N1 = N = 10, D1 = D = 2 and n = 4.
2. Start loop.
3. At i = 1, p = D1/N1 = 0.200, u = 0.37 say. Since u ≥ p,set N1 = 9.
4. At i = 2, p = D1/N1 = 0.222, u = 0.51 say. Since u ≥ p, set N1 = 8.
5. At i = 3, p = D1/N1 = 0.250, u = 0.14 say. Since u < p, set N1 = 7, x = 1, D1 = 1.
6. At i = 4, p = D1/N1 = 0.143, u = 0.84 say. Since u ≥ p, set N1 = 6.
7. End loop.
8. Return x = 1.

Geometric

A variable x is distributed by the geometric distribution when x measures the number of trials to achieve a success, and where the probability of a success, p, remains the same for each trial. The probability of x is listed below.

$$P(x) = p(1-p)^{x-1} \quad x = 1, 2, \ldots$$

The cumulative distribution function of x is the following:

$$F(x) = 1 - (1-p)^x \quad x = 1, 2, \ldots$$

The expected value and the corresponding variance of x are listed below:

$$E(x) = 1/p$$

$$V(x) = (1-p)/p^2$$

To generate a random geometric variate of x (number of trials to achieve a success), the following routine is run:

1. Generate a random continuous uniform $u \sim U(0,1)$.
2. $x = \text{integer}[\ln(1-u)/\ln(1-p)] + 1$, where $\ln = $ natural logarithm.
3. Return x.

Example 5.8 Suppose an experiment is run where the probability of a success is $p = 0.20$ and a random geometric variate of x, the number of trials till the first success, is needed. To accomplish, the following three steps are shown:

1. Generate a random uniform from $u \sim U(0,1)$, say, $u = 0.27$.
2. $x = \text{integer}[\ln(1-.27)/\ln(1-.2)] + 1 = 2$.
3. Return $x = 2$.

When the variable is defined as the number of failures till obtain a success, the variable is $x' = x-1$ and $x' = 0, 1, \ldots$ The probability of x' is below:

$$P(x') = p(1-p)^{x'} \quad x' = 0, 1, 2, \ldots$$

The cumulative distribution function is the following:

$$F(x') = 1 - (1-p)^{x'+1} \quad x' = 0, 1, 2, \ldots$$

Also, $E(x') = E(x)-1 = (1-p)/p$ and $V(x') = V(x) = (1-p)/p^2$.

Pascal

A variable x follows the Pascal distribution when x represents the number of trials needed to gain k successes when the probability of a success is p. This distribution is also called the negative binomial distribution. The probability of x is listed below:

$$P(x) = \binom{x-1}{k-1} p^k (1-p)^{x-k} \qquad x = k, \ k+1, \ \ldots$$

The cumulative distribution function is,

$$F(x) = \sum_{y=k}^{x} P(y) \qquad x = k, \ k+1, \ \ldots$$

The mean and variance of x are given below:

$$E(x) = k/p$$

$$V(x) = k(1-p)/p^2$$

To generate a random Pascal variate of x, the following routine is run:

1. $x = 0$
2. For $i = 1$ to k
   Generate y, a random geometric variate with parameter p
   (Note, $y =$ number of trials till a success.)
   $x = x + y$
   Next i
3. Return x

Example 5.9 Suppose x is distributed as a Pascal variable with $p = 0.5$ and $k = 5$, whereby x represents the number of trials until five successes. The following steps illustrate how x is generated:

1. $x = 0$.
2. At $i = 1$, generate a geometric y with $p = .5$, say $y = 3$. $x = 3$.
3. At $i = 2$, generate a geometric y with $p = .5$, say $y = 1$. $x = 4$.
4. At $i = 3$, generate a geometric y with $p = .5$, say $y = 2$. $x = 6$.
5. At $i = 4$, generate a geometric y with $p = .5$, say $y = 4$ $x = 10$.
6. At $i = 5$, generate a geometric y with $p = .5$, say $y = 2$. $x = 12$.
7. Return $x = 12$.

When the variable is defined as the number of failures till obtain k successes, the variable is $x' = x-k$ and $x' = 0, 1, \ldots$ Also, $E(x') = E(x)-k = k(1-p)/p$ and $V(x') = V(x) = k(1-p)/p^2$.

Poisson

The variable x is described as Poisson distributed when events occur at a constant rate, θ, during a specified interval of time. Could be the number of vehicles crossing an intersection each minute, or the demand for a product over a month's interval of time. The probability of x is listed below.

$$P(x) = \theta^x e^{-\theta}/x! \qquad x = 0, 1, 2, \ldots$$

The expected value and variance of x are shown below:

$$E(x) = \theta$$
$$V(x) = \theta$$

Relation to the Exponential Distribution

The Poisson and the exponential distributions are related since the time, t, between events from a Poisson variable is distributed as exponential with $E(t) = 1/\theta$. This relation is used to randomly generate a value of x.

Generating a Random Poisson Variate

To generate a random variate of x from the Poisson distribution with parameter, θ, the following routine is run.

1. Set $x = 0$, $i = 0$ and $St = 0$.
2. $i = i + 1$, generate a random exponential variate, t, with $E(t) = 1/\theta$, and set $St = St + t$.
3. if $St > 1$, go to step 5.
4. if $St \leq 1$, $x = x + 1$, goto step 2.
5. Return x.

Example 5.10 Assume x is a random variable that follows the Poisson distribution where the expected occurrences in an interval of time is $\theta = 2.4$. In the steps below, note how a random Poisson variate x, is derived from randomly generated exponential t variates with expected value of 1/2.4:

1. $x = 0$ and $St = 0$.
2. At $i = 1$, $t = 0.21$ say, $St = 0.21$, $x = 1$.
3. At $i = 2$, $t = 0.43$ say, $St = 0.64$, $x = 2$.
4. At $i = 3$, $t = 0.09$ say, $St = 0.73$, $x = 3$.
5. At $i = 4$, $t = 0.31$ say, $St = 1.04$.
6. Return $x = 3$.

Example 5.11 A gas station is open 24 h a day where 200–300 vehicles arrive for gas each day, equally distributed. Eighty percent of the vehicles are cars, 15 % are trucks and 5 % motorcycles. The consumption of gas per vehicle is a truncated exponential with a minimum and average known by vehicle type. Cars consume on average 11 gal with a minimum of 3 gal. Trucks consumer a minimum of 8 gal and an average of 20 gal. The motorcycles consume a minimum of 2 gal and an average of 4 gal. The analyst wants to determine the distribution of the total consumption of gas for a day.

A simulation model is developed and run for 1,000 days. On the first day, 228 vehicles enter the station and for each of the 228 vehicles, the type of vehicle and the amount of gas consumed is randomly generated. The sum of gas consumed in the first day is G = 2,596 gal. The simulation is carried on for 1,000 days and the amount of gas consumed is recorded for each day. Now with 1,000 values of G, the gas consumed per day, the next step is to sort the values of G from low to high to yield $G(1) \leq G(2) \leq \ldots \leq G(1000)$. The p-quantile is estimated by $G(p \times 1000)$. For example, the 0.01 quantile is estimated using $0.01 \times 1000 = 10$ where $G(10) = 2,202$, that is the tenth smallest value of G. The table below lists various p-quantiles of the daily gas consumption.

p	G(p)
0.01	2,202
0.05	2,336
0.10	2,436
0.20	2,546
0.30	2,658
0.40	2,768
0.50	2,883
0.60	3,014
0.70	3,138
0.80	3,261
0.90	3,396
0.95	3,503
0.99	3,656

The results show where there is a 5 % chance that G will exceed 3,503 gal and a 1 % chance that G will exceed 3,656 gal. Further, an estimate of the 90 % prediction interval on G becomes $P(2336 \leq G \leq 3503) = 0.90$.

Example 5.12 An always-open truck dealership has ten bays to service trucks for maintenance and service. The arrival rate of trucks is Poisson distributed with 6.67 vehicles per day, and the service rate is also Poisson with a service rate of 1.33 vehicles per day. The vehicles require n parts in the service operation and the probability of n, denoted as P(n), is as follows:

n	3	4	5	6	7	8
P(n)	0.11	0.17	0.22	0.28	0.18	0.04

Four types of parts are described depending on the source of the supplier: PDC (parts distribution center), OEM (original equipment manufacturer), DSH (direct ship supplier), NST (Non stocked part). The table below lists the probability the vehicle needs one of these type of parts, P(type); the service level for each type of part, (SL), where SL = P(part is available in dealer); and the lead time to obtain the part in days (LT). Note, the dealer is limited on space and budget and must set his service levels accordingly. The higher the service level, the more inventory in pieces and in investment.

Part type	PDC	OEM	DSH	NST
P(type)	0.50	0.30	0.19	0.01
SL	0.93	0.92	0.95	0.00
LT	1.25	1.50	2.50	1.50

A simulation model is developed and is run until 5,000 vehicles are processed in the dealership. The first 500 vehicles are used as the transient stage, whereby the equilibrium stage is for the final 4,500 vehicles. This is where all the measurements are tallied. The table below shows some of the statistics gathered from the final 4,500 vehicles.

Bay averages per vehicle:

Empty time	= 0.19 days
Service time	= 0.75 days
Wait time for part(s)	= 0.56 days
Total time	= 1.50 days

Vehicle averages:

Wait time in yard	= 0.09 days
Service time	= 0.75 days
Wait time for part(s)	= 0.56 days
Total time	= 1.40 days

The results show where the average bay is empty for 0.19 days for each vehicle it processes. Further the average service time is 0.75 days and the average idle time per vehicle waiting to receive the out-of-stock part(s) is 0.56 days. The average wait time a vehicle is in the yard prior to service is 0.09 days and the average time in the dealership is 1.40 days. Note also where the average time between vehicles for a bay is 1.50 days.

Summary

This chapter shows how to transform continuous uniform random variates, u ∼ U(0,1), to random discrete variates for a variable that comes from one of the more common discrete probability distributions. The probability distributions described here are the following: discrete arbitrary, discrete uniform, Bernoulli, binomial, hyper-geometric, geometric, Pascal and Poisson.

Chapter 6
Generating Multivariate Random Variates

Introduction

When two or more random variables are jointly related in a probability way, they are labeled as multivariate random variables. The probability of the variables occurring together is defined by a joint probability distribution. In most situations, all of the variables included in the distribution are continuous or all are discrete; and on less situations, they are a mixture between continuous and discrete. This chapter considers some of the more popular multivariate distributions and shows how to generate random variates for each. The probability distributions described here are the following: multivariate discrete arbitrary, multinomial, multivariate hyper geometric, bivariate normal, bivariate lognormal, multivariate normal and multivariate lognormal. The Cholesky decomposition method is also described since it is needed to generate random variates from the multivariate normal and the multivariate lognormal distributions.

Multivariate Discrete Arbitrary

Suppose k discrete random variables $(x_1, \ldots, x_k)$ are jointly related by the probability distribution $P(x_1, \ldots, x_k)$. The sum of the probabilities over all possible values of the k variables is one, i.e.,

$$\sum_{x_1 \ldots x_k} P(x_1 \ldots x_k) = 1.0$$

Consider one of the k variables, say x_j. The marginal probability of x_j, denoted as $P(x_j \ldots)$, is obtained by summing the joint probability distribution over all x_i except x_j, as shown below,

$$P(x_j \ldots) = \sum_{all.x..but.x_j} P(x_1, x_2, \ldots, x_k)$$

N.T. Thomopoulos, *Essentials of Monte Carlo Simulation: Statistical Methods for Building Simulation Models*, DOI 10.1007/978-1-4614-6022-0_6, © Springer Science+Business Media New York 2013

The partial expectation of x_j is obtained as follows,

$$E(x_j \ldots) = \sum_{x_j} x_j P(x_j \ldots)$$

and the partial variance is

$$V(x_j \ldots) = E\left(x_j^2 \ldots\right) - E(x_j \ldots)^2$$

where

$$E\left(x_j^2 \ldots\right) = \sum_{x_j} x_j^2 P(x_j \ldots)$$

Generate a Random Set of Variates

The steps below show how to generate a random set of variates for the variables, $x_1, x_2, \ldots, x_k$.

1. Get $P(x_1 \ldots)$, the marginal probability of x_1, and also $F(x_1 \ldots)$, the corresponding cumulative distribution.

   Generate a random continuous uniform variate $u \sim U(0,1)$.
   Locate the smallest value of x_1 where $u < F(x_1 \ldots)$, say x_{10}.

2. Get $P(x_2|x_{10} \ldots)$, the marginal probability of x_2 given x_{10}, and also $F(x_2|x_{10} \ldots)$, the corresponding cumulative distribution.

   Generate a random uniform continuous variate $u \sim U(0,1)$.
   Locate the smallest value of x_2 where $u < F(x_2|x_{10} \ldots)$, say x_{20}.

3. Get $P(x_3|x_{10}x_{20} \ldots)$, the marginal probability of x_3 given x_{10} and x_{20}, and also $F(x_3|x_{10}x_{20} \ldots)$, the corresponding cumulative distribution.

   Generate a random continuous uniform variate $u \sim U(0,1)$.
   Locate the smallest value of x_3 where $u < F(x_3|x_{10}x_{20} \ldots)$, say x_{30}.

4. Repeat in the same way until get x_{k0}.
5. Return $(x_{10}, \ldots, x_{k0})$.

Example 6.1 Suppose a three variable joint probability distribution with variables, x_1, x_2, x_3 where the possible values for x_1 is 0,1,2, for x_2 it is 0,1, and for x_3 it is 1,2,3. The probability distribution $P(x_1, x_2, x_3)$ is listed below. Note the sum of all probabilities is one. Below shows how to generate one set of random variates.

P(x1,x2,x3)

x3	1	2	3	1	2	3
x2	0			1		
x1						
0	0.12	0.10	0.08	0.08	0.06	0.05
1	0.08	0.06	0.04	0.05	0.04	0.03
2	0.06	0.04	0.02	0.04	0.03	0.02

1. The marginal distribution for x_1 and the associated cumulative distribution is below:

   $P(0..) = 0.49 \ F(0..) = 0.49$
   $P(1..) = 0.30 \ F(1..) = 0.79$
   $P(2..) = 0.21 \ F(2..) = 1.00$
   Generate a random $u \sim U(0,1)$, $u = 0.37$ say. Hence $x_{10} = 0$.

2. The marginal distribution for x_2 given $x_{10} = 0$, and the associated cumulative distribution is below:

   $P(0|0.) = 0.30/0.49 = 0.612 \ F(0|0.) = 0.612$
   $P(1|0.) = 0.19/0.49 = 0.388 \ F(1|0.) = 1.000$
   Generate a random $u \sim U(0,1)$, $u = 0.65$ say. Hence $x_{20} = 1$.

3. The marginal distribution for x_3 given $x_{10} = 0$ and $x_{20} = 1$, and the associated cumulative distribution is below.

   $P(1|01) = 0.08/0.19 = 0.421 \ F(1|01) = 0.421$
   $P(2|01) = 0.06/0.19 = 0.316 \ F(2|01) = 0.737$
   $P(3|01) = 0.05/0.19 = 0.263 \ F(3|01) = 1.000$
   Generate a random $u \sim U(0,1)$, $u = 0.84$ say. Hence $x_{30} = 3$.

4. Return $(x_{10}, x_{20}, x_{30}) = (0, 1, 3)$

Multinomial

Suppose an experiment has k mutually exclusive possible outcomes, $A_1, \ldots, A_k$, with probabilities $p_1, \ldots, p_k$, respectively, and $\sum_{i=1}^{k} p_i = 1.0$. With n independent trials of the experiment, the random variables are $x_1, \ldots, x_k$ representing the number of times event A_i ($i = 1, \ldots, k$) has occurred. Note $\sum_{i=1}^{k} x_i = n$. The probability of x_i is as follows:

$$P(x_1, \ldots, x_k) = n!/[x_1! \ldots x_k!] p_1^{x_1} \cdots p_k^{x_k}$$

The marginal probability of each individual variable, x_i, is a binomial random variable with parameters, n and p_i. The associated mean and variance of x_i are listed below.

$$E(x_i) = np_i$$

$$V(x_i) = np_i(1 - p_i)$$

Generating Random Multinomial Variates

The steps below show how to randomly generate a set of multinomial variates $(x_1, \ldots, x_k)$ from n trials with probabilities, $p_1, \ldots, p_k$.

1. For i = 1 to k

$$p_i' = p_i / \sum_{j=i}^{k} p_j$$

$$n_i' = n - \sum_{j=1}^{i-1} x_j$$

Generate a random Binomial variate, x_i, with parameters, n_i' and p_i'.

Next i

2. Return $x_1, \ldots, x_k$

Example 6.2 Suppose an experiment with three possible outcomes with probabilities 0.5, 0.3 and 0.2, respectively, where five trials of the experiment are run. Of need is to randomly generate the multinomial variate set (x_1, x_2, x_3) for this situation. The steps below show how this is done.

1. At i = 1, with parameters $n_1' = 5$, $p_1' = 0.5$, generate a random binomial, say $x_1 = 2$.
2. At i = 2, with parameters $n_2' = 3$, $p_2' = 0.6$, generate a random binomial, say $x_2 = 2$.
3. At i = 3, with parameters $n_3' = 1$, $p_3' = 1.0$, generate a random binomial, say $x_3 = 1$.
4. Return $(x_1, x_2, x_3) = (2, 2, 1)$.

Multivariate Hyper Geometric

Suppose a population of N items where some are non-defective and the remainder are defective falling into k defective categories with number of defectives, $D_1, \ldots, D_k$. A sample of n items are taken without replacement and the outcomes are $x_1, \ldots, x_k$

defective items. Note, x_i = number of defective items of the ith category in the sample. The random variables follow the Multivariate Hyper Geometric distribution. The probability distribution is listed below,

$$P(x_1, \ldots, x_k) = \binom{N - SD}{n - Sx} \binom{D_1}{x_1} \cdots \binom{D_k}{x_k} \Big/ \binom{N}{n}$$

where

$$SD = \sum_{i=1}^{k} D_i = \text{sum of defective items in the population}$$

$$Sx = \sum_{i=1}^{k} x_i = \text{sum of defective items in the sample}$$

$n - Sx$ = sum of non-defective items in the sample

Generating Random Variates

To generate a random set of output variates for the Multivariate Hyper Geometric distribution, the following routine is run. Recall the notation of $N, n, D_1, \ldots, D_k$ and $x_1, \ldots, x_k$.

1. Initialize $D1_i = D_i$ for $i = 1$ to k, and $N1 = N$.
2. For $j = 1$ to n

   Generate a random uniform $u \sim U(0,1)$.
   $F = 0$
   For $i = 1$ to k

   $p = D1_i/N1$
   $F = F + p$

   If $u < F$, $x_i = x_i + 1$, $D1_i = D1_i - 1$, go to 3

   Next i

   3. $N1 = N1 - 1$

   Next j

4. Return $(x_1, \ldots, x_k)$.

Example 6.3 Suppose a lot of size 20 comes in to a receiving dock with three types of defectives. There are 4 defective of type 1, 3 defectives of type 2, and 2 defectives of type 3. Eleven of the items have no defectives. A sample of size four is taken without replacement and of need is to generate a random set of output variates, x_1, x_2, x_3. The method to obtain the variates is shown below. For simplicity, the fractions are carried out only to two decimal places.

1. Initialize N1 = 20, $(D1_1, D1_2, D1_3) = (4,3,2)$,

   $x_1 = 0, x_2 = 0, x_3 = 0$.

2. At $j = 1, u = 0.71$ say, F = 0.

   At $i = 1, p = 4/20 = 0.20, F = 0.20$
   At $i = 2, p = 3/20 = 0.15, F = 0.35$
   At $i = 3, p = 2/20 = 0.10, F = 0.45$
   N1 = 19

3. At $j = 2, u = 0.34$ say, F = 0.

   At $i = 1, p = 4/19 = 0.21, F = 0.21$
   At $i = 2, p = 3/19 = 0.16, F = 0.37, x_2 = x_2 + 1, D1_2 = 2$
   N1 = 18

4. At $j = 3, u = 0.63$ say, F = 0.

   At $i = 1, p = 4/18 = 0.22, F = 0.22$
   At $i = 2, p = 2/18 = 0.11, F = 0.33$
   At $i = 3, p = 2/18 = 0.11, F = 0.44$
   N1 = 17

5. At $j = 4, u = 0.14$ say, F = 0.

   At $i = 1, p = 4/17 = 0.23, F = 0.23, x_1 = x_1 + 1, D1_1 = 3$
   N1 = 16

6. Return $(x_1, x_2, x_3) = (1,1,0)$

Bivariate Normal

Consider variables x_1 and x_2 that are jointly related via the bivariate normal distribution (BVN) as below:

$$f(x_1, x_2) = 1/[2\pi\sigma_1\sigma_2\sqrt{(1-\rho 2)}\,] \exp\{-\{[(x_1 - \mu_1)/\sigma_1]^2 + [(x_2 - \mu_2)/\sigma_2]^2 - 2\rho[(x_1 - \mu_1)(x_2 - \mu_2)/\sigma_1\sigma_2]\}/2(1-\rho^2)\}$$

where $(\mu_1, \mu_2, \sigma_1, \sigma_2, \rho)$ are five parameters of the distribution.

Marginal Distributions

The marginal distributions of x_1 and x_2 are normally distributed, whereby,

$$x_1 \sim N(\mu_1, \sigma_1^2)$$

and

$$x_2 \sim N(\mu_2, \sigma_2{}^2)$$

The expected value and variance of x_1 are,

$$E(x_1) = \mu_1$$

$$V(x_1) = \sigma_1{}^2$$

respectively.

The corresponding values for x_2 are

$$E(x_2) = \mu_2,$$

$$V(x_2) = \sigma_2{}^2$$

The correlation between x_1 and x_2 is ρ. Note,

$$\rho = \sigma_{12}/(\sigma_1\sigma_2)$$

where σ_{12} is the covariance between x_1 and x_2. The covariance is also denoted as $C(x_1,x_2)$ and is obtained from,

$$C(x_1, x_2) = E(x_1 x_2) - E(x_1)E(x_2)$$

Conditional Distributions

When $x_1 = x_{10}$, say, the conditional mean of x_2 is

$$\mu_{x2|x10} = \mu_2 + \rho(\sigma_2/\sigma_1)(x_{10} - \mu_1)$$

The corresponding variance is

$$\sigma_{x2|x1}{}^2 = \sigma_2{}^2(1 - \rho^2)$$

and is the same for all values of x_1.

The associated conditional distribution of x_2 given x_{10} is also normally distributed as,

$$x_2|x_{10} \sim N\left(\mu_{x2|x10}, \sigma_{x2|x1}{}^2\right)$$

In the same way, when $x_2 = x_{20}$, say, the conditional mean of x_1 is

$$\mu_{x1|x20} = \mu_1 + \rho(\sigma_1/\sigma_2)(x_{20} - \mu_2)$$

The corresponding variance is

$$\sigma_{x1|x2}^2 = \sigma_1^2(1 - \rho^2)$$

and is the same for all values of x_2.

The associated conditional distribution of x_1 given x_{20} is also normally distributed as,

$$x_1|x_{20} \sim N\left(\mu_{x1|x20}, \sigma_{x1|x2}^2\right)$$

Generate Random Variates (x_1, x_2)

To generate a random pair of x_1 and x_2 with parameters (μ_1, μ_2, σ_1, σ_2, ρ), the following routine is run.

1. Generate a random standard normal, $z \sim N(0,1)$, say z_1.
2. A random x_1 is computed by $x_{10} = \mu_1 + z_1\sigma_1$
3. The conditional mean and variance of x_2 now become, $\mu_{x2|x10} = \mu_2 + \rho(\sigma_2/\sigma_1)$ $(x_{10} - \mu_1)$ and $\sigma_{x2|x1}^2 = \sigma_2^2(1 - \rho^2)$, respectively.
4. Generate another random standard normal, $z \sim N(0,1)$, say z_2.
5. The random x_2 is computed by $x_{20} = \mu_{x2|x10} + z_2\sigma_{x2|x1}$
6. Return (x_{10}, x_{20})

Example 6.4 Suppose the pair (x_1, x_2) are related by the bivariate normal distribution with parameters (5,8, 1,4,0.5) and a random variate of the pair is needed. The following routine is run.

1. Generate a random standard normal from $z \sim N(0,1)$, say, $z_1 = 0.72$.
2. The random variate of x_1 becomes $x_{10} = 5 + 0.72(1) = 5.72$
3. The mean and variance of the conditional x_2 variable now are, $\mu_{x2|5.72} = 8 + 0.5$ $(2/1)(5.72 - 5.00) = 8.72$, and $\sigma_{x2|5.72}^2 = 4(1 - 0.5^2) = 3$
4. The conditional standard deviation is $\sigma_{x2|5.72} = 3^{0.5} = 1.732$
5. Generate another standard normal variate of $z \sim N(0,1)$, say, $z_2 = -1.08$.
6. The random variate for x_2 now becomes, $x_{20} = 8.72 - 1.08 (1.732) = 6.85$
7. Return (x_{10}, x_{20}) = (5.72, 6.85).

Example 6.5 Consider the bivariate normal variables (x_1, x_2) with parameters $\mu_1 = 0$, $\mu_2 = 0$, $\sigma_1 = 1$, $\sigma_2 = 1$ and ρ. A researcher is seeking the four following cumulative probability distributions, $F(0,0)$, $F(1,0)$, $F(0,1)$, $F(1,1)$, at the three correlations, $\rho = -0.5$, 0.0 and 0.5. Recall, $F(x_{10}, x_{20}) = P(x_1 \leq x_{10}$, and $x_2 \leq x_{20})$. To comply, a simulation model is developed to find the probabilities needed. For a given x_{10}, and x_{20}, n trials are run where the values of x_1 and x_2 are randomly generated to conform with the stated parameters. The program is run with n trials where g represents the number of times in the n trials the generated values were both less or equal to x_{10}, and x_{20}. The estimate of the probability is computed by $F(x_{10}, x_{20}) = g/n$.

The table below lists the simulation findings where various number of trials are run at n = 50, 100, 500 and 1,000. Note, at $\rho = 0$, the true value of the cumulative distribution is known at x_{10}, and x_{20} and is listed in the table at n = ∞. The results point to the need for more trials to sharpen the probability estimates.

ρ	n	F(0,0)	F(1,0)	F(0,1)	F(1,1)
−0.50	50	0.180	0.310	0.460	0.760
	100	0.160	0.410	0.340	0.710
	500	0.174	0.372	0.372	0.696
	1,000	0.159	0.360	0.357	0.652
0.00	50	0.360	0.580	0.480	0.780
	100	0.250	0.490	0.330	0.650
	500	0.234	0.412	0.400	0.668
	1,000	0.233	0.395	0.405	0.687
	∞	0.250	0.421	0.421	0.708
0.50	50	0.380	0.500	0.520	0.760
	100	0.270	0.430	0.420	0.660
	500	0.326	0.476	0.448	0.742
	1,000	0.370	0.468	0.509	0.748

Bivariate Lognormal

When the pair x_1, x_2 are bivariate lognormal (BVLN), the distribution is noted as BVLN(μ_{y1}, μ_{y2}, σ_{y1}, σ_{y2}, ρ_y), where μ_{y1} and μ_{y2} are the means of the $y_1 = \ln(x_1)$ and $y_2 = \ln(x_2)$. Also σ_{y1} and σ_{y2} are the corresponding standard deviations of y_1 and y_2. ρ_y is the correlation between y_1 and y_2. The transformed pair, (y_1, y_2) are distributed by the bivariate normal distribution and the notation is BVN(μ_{y1}, μ_{y2}, σ_{y1}, σ_{y2}, ρ_y). Note $x_1 = e^{y_1}$ and $x_2 = e^{y_2}$.

Generate a Random Pair (x_1, x_2)

To generate a random pair (x_1, x_2), the following routine is run.

1. Find μ_{y1}, μ_{y2}, σ_{y1}, σ_{y2} and ρ_y.
2. From the standard normal, $z \sim N(0,1)$, generate two random values, say, z_1, z_2.
3. Get a random y_1 by $y_{10} = \mu_{y1} + z_1 \sigma_{y1}$

4. Now find the conditional mean and standard deviation of y_2, given y_{10}, from $\mu_{y2|y10} = \mu_{y2} + \rho_y(\sigma_{y2}/\sigma_{y1})(y_{10} - \mu_{y1})$, and $\sigma_{y2|y1}^2 = \sigma_{y2}^2(1 - \rho_y^2)$.
5. Get a random y_2 by $y_{20} = \mu_{y2|y10} + z_2\,\sigma_{y2|y1}$.
6. Now, $x_1 = e^{y10}$ and $x_2 = e^{y20}$.
7. Return x_1, x_2.

Example 6.6 Assume x_1, x_2 are a pair of bivariate lognormal variables with parameters BVLN(5,8,1,2,0.5), and a set of random variates is needed. The following steps are followed:

1. From $z \sim N(0,1)$, get the pair of random variates, say: $z_1 = 0.72$ and $z_2 = -1.08$.
2. A random y_1 becomes $y_{10} = 5 + 0.72(1) = 5.72$
3. The mean and standard deviation of y_2 given $y_{10} = 5.72$ are $\mu_{y2|5.72} = 8.72$, and $\sigma_{y2|5.72} = 1.732$.
4. So now, the random y_2 becomes, $y_{20} = 8.72 - 1.08(1.732) = 6.85$.
5. Finally, the pair needed are $x_1 = e^{5.72} = 304.90$ and $x_2 = e^{6.85} = 943.88$.
6. Return $(x_1, x_2) = (304.90, 943.88)$.

Multivariate Normal

When k variables, $x_1, \ldots, x_k$, are related by the multivariate normal distribution, the parameters are μ and Σ. The parameter μ is a k-dimensional vector whose transpose $\mu^T = [\mu_1, \ldots, \mu_k]$ houses the mean of each of the variables, and Σ is a k-by-k matrix, that contains σ_{ij} in row i and column j, where σ_{ij} is the covariance between variables i and j. Note, the covariances along the main diagonal, σ_{ii}, are the variances of variable i, i = 1 to k. Thus $\sigma_i^2 = \sigma_{ii}$, i = 1 to k.

Cholesky Decomposition

An important relation is the Cholesky decomposition of the matrix Σ where,

$$\Sigma = CC^T$$

and C is a k-by-k matrix where the upper diagonal is all zeros and the diagonal and lower diagonal contain the elements c_{ij}. For a full discussion, see Gentle (1998).

The values of the elements of matrix C are computed in the three stages listed below.

Column 1 elements: $c_{i1} = \sigma_{i1}/\sqrt{\sigma_{11}}$ i = 1, ..., k

Main diagonal elements: $c_{ii} = \sqrt{\sigma_{ii} - \sum_{m=1}^{i-1} c_{im}^2}$ i = 2, ..., k

Lower diagonal elements: $c_{ij} = \left[\sigma_{ij} - \sum\limits_{m=1}^{j-1} c_{im}c_{jm}\right]/cjj \qquad 1<j<i \leq k$

As stated earlier, the upper diagonal elements of C are all zero.

Generate a Random Set $[x_1, \ldots, x_k]$

To generate a random set of variables from the k-dimensional multivariate distribution, the following routine is run.

1. From the variance-covariance matrix Σ, compute the matrix C.
2. Using the standard normal distribution, $z \sim N(0,1)$, generate k random variates: $z_1, \ldots, z_k$, and insert them in a k-dimensional vector Z, where the transpose is $Z^T = [z_1, \ldots, z_k]$.
3. The random variates of the k variables will be placed in a k-dimensional vector X, whose transpose is $X^T = [x_1, \ldots, x_k]$.
4. So now, the k random variates of the k variables are obtained by the following matrix manipulation, $X = CZ + \mu$.
   Another way is to compute the x_i as below:

$$x_i = \sum_{j=1}^{k} c_{ij}z_j + \mu_i \qquad i = 1, \ldots, k$$

Example 6.7 When $k = 2$, the matrices, Σ, C and X are as below:

$$\Sigma = \begin{bmatrix} \sigma_{11} & \sigma_{12} \\ \sigma_{21} & \sigma_{22} \end{bmatrix} = \begin{bmatrix} \sigma_1^2 & \rho\sigma_1\sigma_2 \\ \rho\sigma_2\sigma_1 & \sigma_2^2 \end{bmatrix}$$

$$C = \begin{bmatrix} c_{11} & c_{12} \\ c_{21} & c_{22} \end{bmatrix} = \begin{bmatrix} \sigma_1 & 0 \\ \rho\sigma_2 & \sigma_2\sqrt{(1-\rho^2)} \end{bmatrix}$$

$$X = \begin{bmatrix} x_1 \\ x_2 \end{bmatrix} = \begin{bmatrix} \mu_1 + z_1\sigma_1 \\ \mu_2 + z_1\rho\sigma_2 + z_2\sigma_2\sqrt{(1-\rho^2)} \end{bmatrix}$$

Example 6.8 Consider the $k = 3$ dimensional multivariate normal distribution with the mean vector and variance covariance matrix as below.

$$\mu = \begin{bmatrix} 100 \\ 80 \\ 140 \end{bmatrix}$$

$$\Sigma = \begin{bmatrix} 64 & 20 & -10 \\ 20 & 25 & 36 \\ -10 & 36 & 100 \end{bmatrix}$$

The C matrix becomes:

$$
C = \begin{bmatrix} 8 & 0 & 0 \\ 2.5 & 4.33 & 0 \\ -1.25 & 9.04 & 4.09 \end{bmatrix}
$$

When the three entries of the standard normal variates are: $z_1 = -0.02$, $z_2 = 2.00$, $z_3 = -1.20$, and the matrix manipulation of $X = CZ + \mu$ is applied, the random normal variates become:

$$
X = \begin{bmatrix} x_1 \\ x_2 \\ x_3 \end{bmatrix} = \begin{bmatrix} 99.84 \\ 88.16 \\ 153.20 \end{bmatrix}
$$

Multivariate Lognormal

When k variables, $x_1, \ldots, x_k$, are related by the multivariate lognormal distribution, the associated bivariate normal variables are $y_1, \ldots, y_k$, where $y_i = \ln(x_i)\, i = 1, \ldots, k$. The parameters for this distribution are derived from the k variables, $y_1, \ldots, y_k$. The parameters are listed in the transposed k-dimensional vector, μ, whose transpose is $\mu^T = [\mu_1, \ldots, \mu_k]$, that houses the mean of each of the y_i variables, and the k-by-k matrix, Σ, that contains σ_{ij} in row i and column j, where σ_{ij} is the covariance between variables y_i and y_j. Note, the covariances along the main diagonal, σ_{ii}, are the variances of variable y_i, $i = 1$ to k. Thus $\sigma_i^2 = \sigma_{ii}$, $i = 1$ to k.

Cholesky Decomposition

The Cholesky decomposition of the matrix Σ is used here where,

$$
\Sigma = CC^T
$$

and C is a k-by-k matrix where the upper diagonal is all zeros and the diagonal and lower diagonal contain the elements c_{ij}. As shown earlier, the values of the elements of matrix C are computed in the three stages listed below.

Column 1 elements: $c_{i1} = \sigma_{i1}/\sqrt{\sigma_{11}}$ $\qquad\qquad$ $i = 1, \ldots, k$

Main diagonal elements: $c_{ii} = \sqrt{\sigma_{ii} - \sum_{m=1}^{i-1} c_{im}^2}$ $\quad i = 2, \ldots, k$

Lower diagonal elements: $c_{ij} = \left[\sigma_{ij} - \sum_{m=1}^{j-1} c_{im}c_{jm}\right]/c_{jj}$ $\quad 1 < j < i \le k$

The upper diagonal elements of C are all zero.

Generate a Random Set [x₁, ... , xₖ]

To generate a random set of variables from the k-dimensional multivariate distribution, the following routine is run.

1. Convert the lognormal variables x to y by $y_i = \ln(x_i)$ $i = 1, \ldots, k$.
2. From the data of $(y_1, \ldots, y_k)$, compute the mean and variance-covariance matrix Σ.
3. From Σ compute the matrix C.
4. Using the standard normal distribution, $z \sim N(0,1)$, generate k random variates: $z_1, \ldots, z_k$, and insert them in a k-dimensional vector Z, where the transpose is $Z^T = [z_1, \ldots, z_k]$.
5. The random variates of the k variables will be placed in a k-dimensional vector Y, whose transpose is $Y^T = [y_1, \ldots, y_k]$.
6. So now, the k random variates of the k variables are obtained by the following matrix manipulation, $Y = CZ + \mu$.
   Another way is to compute the y_i as below:

$$y_i = \sum_{j=1}^{k} c_{ij}z_j + \mu_i \qquad i = 1, \ldots, k$$

7. Finally, convert the k random y variates to k random x variates by the relation $x_i = e^{y_i}$, $i = 1, \ldots, k$.
8. Return $(x_1, \ldots, x_k)$

Example 6.9 Consider variables (x_1, x_2, x_3) from the multivariate lognormal distribution with converted variables (y_1, y_2, y_3) from the associated multivariate normal distribution, and their parameters below:

$$\mu_y = \begin{bmatrix} -5 \\ 2 \\ -8 \end{bmatrix}$$

$$\Sigma_y = \begin{bmatrix} 64 & 20 & -10 \\ 20 & 25 & 36 \\ -10 & 36 & 100 \end{bmatrix}$$

The C matrix from the above becomes:

$$C = \begin{bmatrix} 8 & 0 & 0 \\ 2.5 & 4.33 & 0 \\ -1.25 & 9.04 & 4.09 \end{bmatrix}$$

When the three entries of the standard normal variates are: $z_1 = 0.72$, $z_2 = 1.00$, $z_3 = -1.04$, and the matrix manipulation of $Y = CZ + \mu_y$ is applied, the random normal variates become:

$$Y = \begin{bmatrix} y_1 \\ y_2 \\ y_3 \end{bmatrix} = \begin{bmatrix} 0.76 \\ 8.13 \\ -4.11 \end{bmatrix}$$

Finally, convert y_1, y_2, y_3, to x_1, x_2, x_3 by $x_i = e^{y_i}$ for $i = 1, 2, 3$. The multivariate lognormal variates are below.

$$X = \begin{bmatrix} x_1 \\ x_2 \\ x_3 \end{bmatrix} = \begin{bmatrix} 1.82 \\ 3394.80 \\ 0.02 \end{bmatrix}$$

Summary

This chapter considers some of the more popular multivariate distributions and shows how to generate random variates for each. The probability distributions described are the following: multivariate discrete arbitrary, multinomial, multivariate hyper geometric, bivariate normal, bivariate lognormal, multivariate normal and multivariate lognormal. The Cholesky decomposition method is also presented because of its important role in generating random variates from the multivariate normal and multivariate lognormal distributions.

Chapter 7
Special Applications

Introduction

This chapter shows how to generate random variates for applications that are not directly bound by a probability distribution as was described in some of the earlier chapters. The applications are instructively useful and often are needed as such in simulation models. They are the following: Poisson process, constant Poisson process, batch arrivals, active redundancy, standby redundancy, random integers without replacement and poker.

Poisson Process

There are many simulation models where a series of events take place over a fixed time horizon, say, from t = 0 to T. In this section, the arrivals are from a Poisson process, whereby the time between events are distributed via the exponential density. Further, when the expected time between arrivals changes over the time horizon, additional information is needed concerning various points of time in the interval, B(j), and the associated expected time between arrivals, A(j). At t = 0, the average time between arrivals is A(1), and the point in time is B(1) = 0, An interval of time later, say at B(2), the average time between arrivals is A(2), and so forth. In this way, A(j) and B(j) jointly identify how the average time between arrivals vary from t = 0 to T. At t = T, the last entry of j occurs and is denoted as j = J, and B (J) = T. Note, A(J) gives the associated average time at t = T, whereby B(J) = T. For any other time from 0 to T, interpolation is used to determine the average time at t, as is described in the routine below:

1. Parameters and initial values: T = length of time horizon, J = number of points in time from 0 to T, B(j) = the j-th point in time, and A(j) = average time between arrivals at time B(j), j = 1 to J, t(0) = 0 is the starting point, and n is an index.

N.T. Thomopoulos, *Essentials of Monte Carlo Simulation: Statistical Methods for Building Simulation Models*, DOI 10.1007/978-1-4614-6022-0_7, © Springer Science+Business Media New York 2013

2. Using t(n) and {B(j) j = 1 to J}, find the minimum jo where t(n) ≥ B(jo).
3. The average time between arrivals becomes: A = A(jo) + {[t(n)−B(jo)]/ [B(jo + 1)−B(jo)]}[A(jo + 1)−A(jo)].
4. Get a random uniform u ~ U(0,1).
5. Use A and u and the exponential density to generate the random time between the arrivals, x = −A × ln(1 − u).
6. If [t(n−1) + x] ≤ T, n = n + 1, t(n) = [t(n−1) + x], go to 2.
7. If [t(n−1) + x] > T, end, go to 8.
8. Return {t(i), i = 1 to n}.

Example 7.1 Suppose T = 24 and J = 5. {B(j), j = 1 to 5} = {0, 8, 14, 18, 24}, {A(j), j = 1 to 5} = {5, 4, 2, 3, 5}, and the time between arrivals are from a Poisson process, with the exponential density. The routine below shows how to find the time of arrival for the first three units:

1. n = 0.
   At t(0) = 0, jo = 1, since B(1) ≤ t(0) < B(2).
   The average time is: A = 5 + (0−0)/(8−0) × (4−5) = 5.0.
   Get u ~ U(0,1), assume u = 0.72.
   The random time is: x = −5.0 × ln(1−0.72) = 6.36.
   n = n + 1 = 1, t(1) = 0 + 6.36 = 6.36.
2. n = 1.
   At t(1) = 6.36, jo = 1, since B(1) ≤ t(1)<B(2).
   The average time is: A = 5 + (6.36 − 0)/(8 − 0) × (4 − 5) = 4.205.
   Get u ~ U(0,1), assume u = 0.38.
   The random time is: x = −4.205 × ln(1 − 0.38) = 2.01.
   n = n + 1 = 2, t(2) = 6.36 + 2.01 = 8.37.
3. n = 2.
   At t(2) = 8.37, jo = 2, since B(2) ≤ t(2)<B(3).
   The average time is: A = 4 + (8.37 − 8)/(14 − 8) × (2 − 4) = 3.666.
   Get u ~ U(0,1), assume u = 0.17.
   The random time is: x = 3.666 × ln(1 − 0.17) = 0.68.
   n = n + 1 = 3, t(3) = 8.37 + 0.68 = 9.05.
4. The first three arrivals occur at times: 6.36, 8.37, 9.05.

Constant Poisson Process

In the event the expected inter-arrival time is the same for the whole time horizon, then J = 2 periods in the time horizon, B(1) = 0 is the start time, B(2) = T is the end time, and A(1) = A(2) are the arrival rates for time periods 1 and 2.

Batch Arrivals

Consider a simulation model where units arrive to a system in batch sizes of one or more. The model generates the random time of arrival and the associated batch size. One way to describe the batch size distribution is by the modified Poisson distribution. Since each individual batch size, x, is one or larger, the expected value of x is $E(x) \geq 1.0$. The modified Poisson becomes $x = y + 1$, where y is a Poisson variable with mean $\mu = E(x)-1$. So, to generate a random x, the following routine is run:

1. For a Poisson variable y with parameter μ, generate a random Poisson y.
2. Set $x = y + 1$.
3. Return x.

Example 7.2 Suppose the average batch size for a simulation run is 1.6, and a random variate of the batch size, y, is needed. The following routine is run:

1. From the Poisson distribution with parameter $\mu = E(x)-1 = 0.6$, generate a random y. Assume the random Poisson $y = 0$.
2. Hence, $x = y + 1 = 0 + 1 = 1$.
3. Return $x = 1$.

Active Redundancy

An active redundancy is when the reliability of a unit, with m components, is satisfied as long as one of the components of the unit is still running. The m components are run simultaneously and the last component to fail is the run time of the unit. Assume the run time, y, of each component is based on the exponential distribution with parameter θ. The run times for the m components are $(y_1, \ldots, y_m)$ and the run time for the unit becomes, $x = \max(y_1, \ldots, y_m)$.

Generate a Random Variate

To generate a random variate for a unit with an active redundancy of m components with expected run time $E(y) = 1/\theta$, the following routine is run:

1. For $i = 1$ to m.
2. From the continuous uniform $u \sim U(0,1)$, generate a random u.
3. Generate a random exponential by $y_i = (-1/\theta) \ln(1 - u)$.
4. Next i.
5. $x = \max(y_1, \ldots, y_m)$.
6. Return x.

Example 7.3 Consider a unit that has three active redundant components with run times following the exponential density and each with an expected run time of 10 h. The routine below generates one random variate of the unit run time:

1. At $i = 1$, generate a random exponential with mean 10, say, $y_1 = 7.4$.
2. At $i = 2$, generate a random exponential with mean 10, say, $y_2 = 15.1$.
3. At $i = 3$, generate a random exponential with mean 10, say, $y_3 = 4.2$.
4. The run time for the unit is $x = \max(7.4, 15.1, 4.2) = 15.1$.
5. Return $x = 15.1$.

Example 7.4 A component is in the design stage and will include m identical subcomponents in an active redundancy manner. All m subcomponents start and run together. The component run time ends when the last subcomponent fails. The time to fail, t, for each subcomponent follows a gamma distribution with parameters $k = 3$ and $1/\theta = 6.0$ (1,000 h), whereby the time to fail, t, has a mean of $E(t) = 18$ (1,000 h). The design engineer wants to know the minimum number of the subcomponents to include in the active redundancy package so that the time to fail for the component, T, has a reliability of R equal to 0.99 or greater at 20 (1,000 h). Note $T = \max(t_1, \ldots, t_m)$ and the goal is to have $R = P[T \geq 20 \ (1000 \ \text{hours})] \geq 0.99$.

A simulation model is developed to find the minimum number of subcomponents to achieve the reliability specified. In the table, m denotes the number of subcomponents, n is the number of trials in a run – where in each trial, T is computed from m random variates of the gamma distribution with the stated parameters, g is number of the trials in the run where $T \geq 20$ (1,000 h), and $R = g/n$ is an estimate of the probability that T will be 20 (1,000 h) or greater. The simulation results are listed below where the number of components (m) increases by one with each simulation run of $n = 1,000$ trials. Note, at $m = 11$, $R = 0.993$ and at $m = 12$, $R = 0.998$. Hence, the minimum value of m is 11 and the reliability is estimated as $R = 0.993$.

m	n	g	R
1	1,000	318	0.318
2	1,000	588	0.588
3	1,000	707	0.707
4	1,000	802	0.802
5	1,000	883	0.883
6	1,000	927	0.927
7	1,000	961	0.961
8	1,000	975	0.975
9	1,000	973	0.973
10	1,000	980	0.980
11	1,000	993	0.993
12	1,000	998	0.998

Standby Redundancy

A unit with standby redundancy is defined when the run time of the unit is the sum of the run times of m components that are run one after the other. That is, when one component fails, another starts running. This model assumes the run time, y, of each component is based on the exponential distribution with parameter θ. The run times for the m components are $(y_1, \ldots, y_m)$, and the run time for the unit becomes, $x = (y_1 + \ldots + y_m)$.

Generate a Random Variate

To generate a random variate for a unit with a standby redundancy of m components with expected run time $E(y) = 1/\theta$, the following routine is run:

1. For $i = 1$ to m.
2. From the continuous uniform $u \sim U(0,1)$, generate a random u.
3. Generate a random exponential by $y_i = (-1/\theta) \ln(1 - u)$.
4. Next i.
5. $x = (y_1 + \ldots + y_m)$.
6. Return x.

Example 7.5 Consider a unit that has four standby redundant components with run times following the exponential density and each with an expected run time of 5 h. The routine below generates one random variate of the unit run time:

1. At $i = 1$, generate a random exponential with mean 5, say, $y_1 = 2.7$.
2. At $i = 2$, generate a random exponential with mean 5, say, $y_2 = 9.3$.
3. At $i = 3$, generate a random exponential with mean 5, say, $y_3 = 7.2$.
4. At $i = 4$, generate a random exponential with mean 5, say, $y_4 = 1.8$.
5. The run time for the unit is $x = (2.7 + 9.3 + 7.2 + 1.8) = 19.0$.
6. Return $x = 19.0$.

Example 7.6 A component is in the design stage and will include m identical subcomponents in a standby redundancy manner. One subcomponent is run at a time; when it fails, and another is still available, the next subcomponent starts its run. The component run time ends when the last subcomponent fails. The time to fail for each subcomponent follows a gamma distribution with parameters $k = 3$ and $1/\theta = 6.0$ (1,000 h), whereby the time to fail, t, has a mean of $E(t) = 18$ (1,000 h). The design engineer wants to know the minimum number of the sub-components to include in the standby redundancy package so that the time to fail for the component, T, has a reliability of R equal to 0.99 or larger at 20 (1,000 h). Note $T = t_1 + \ldots + t_m$ and the goal is to have $R = P[T \geq 20 \ (1000 \ \text{hours})] \geq 0.99$.

A simulation model is developed to find the minimum number of subcomponents to achieve the reliability specified. In the table, m denotes the number of subcomponents, n is the number of trials in a run – where in each trial T is computed from m random variates of the gamma distribution with the stated parameters, g is number of the trials in the run where $T \geq 20$ (1,000 h), and $R = g/n$ is an estimate of the probability that T will be 20 (1,000 h) or greater. The simulation results are listed below where the number of components (m) increases by one with each simulation run of n = 1,000 trials. Note, at m = 3, R = 0.993 and at m = 4, R = 0.999. Hence, the minimum value of m is three and the reliability is estimated as R = 0.993.

m	n	g	R
1	1,000	348	0.348
2	1,000	871	0.871
3	1,000	993	0.993
4	1,000	999	0.999

Random Integers Without Replacement

Consider N unique items where n of them will be arranged in a random sequence and none will be repeated. The N items are identified by D(i) i = 1 to N. The n items in sequence are E(1), ..., E(n).

Generate a Random Sequence

The routine below shows how to generate a random sequence of n samples without replacement from N unique items in a population:

1. The parameters are N = population size, n = sample size without replacement, {D(i), i = 1 to N} identifies the N unique items, and {E(j) j = 1 to n} denotes the n random items in sequence.
2. ND = N = number of unique items remaining.
3. For j = 1 to n.
4. Generate a random discrete uniform integer, k, (1−ND).
   Set E(j) = D(k).
5. ND = ND−1.
6. For m = k to ND.
7. D(m) = D(m + 1).
8. Next m.
9. Next j.
10. Return [E(1), ..., E(n)].

Example 7.7 Suppose a population of N = 10 integers, 1, 2, 3, 4, 5, 6, 7, 8, 9, 10, where n = 5 will be selected randomly in sequence. The routine below is run:

1. At j = 1, ND = 10 and D(i) = (1,2,3,4,5,6,7,8,9,10).
   Using the discrete uniform (1,10), randomly generate k, say k = 7.

Hence, E(1) = 7. ND = 9 and D(i) = (1,2,3,4,5,6,8,9,10).
2. At j = 2, ND = 9 and D(i) = (1,2,3,4,5,6,8,9,10).
   Using the discrete uniform (1,9), randomly generate k, say k = 4.
   Hence, E(2) = 4. ND = 8 and D(i) = (1,2,3,5,6,8,9,10).
3. At j = 3, ND = 8 and D(i) = (1,2,3,5,6,8,9,10).
   Using the discrete uniform (1,8), randomly generate k, say k = 6.
   Hence, E(3) = 8. ND = 7 and D(i) = (1,2,3,5,6,9,10).
4. At j = 4, ND = 7 and D(i) = (1,2,3,5,6,9,10).
   Using the discrete uniform (1,7), randomly generate k, say k = 4.
   Hence, E(4) = 5. ND = 6 and D(i) = (1,2,3,6,9,10).
5. At j = 5, ND = 6 and D(i) = (1,2,3,6,9,10).
   Using the discrete uniform (1,6), randomly generate k, say k = 6.
   Hence, E(5) = 10. ND = 5 and D(i) = (1,2,3,6,9).
6. Return (7, 4, 8, 5, 10).

Poker

A deck of cards has 52 unique items. For simplicity, the items are here labeled as
D(i) i = 1 to 52. The notation is (H, D, S, C) for (hearts, diamonds, spades, clubs)
and (A,2,3,4,5,6,7,8,9,10,J,Q,K) for (ace, 2,3,4,5,6,7,8,9,10, jack, queen, king). The
52 items become:

{D(i) i = 1 to 13} = {AH, 2H, 3H, 4H, 5H, 6H, 7H, 8H, 9H, 10H, JH, QH, KH}
{D(i) i = 14 to 26} = {AD, 2D, 3D, 4D, 5D, 6D, 7D, 8D, 9D, 10D, JD, QD, KD}
{D(i) i = 27 to 39} = {AS, 2S, 3S, 4S, 5S, 6S, 7S, 8S, 9S, 10S, JS, QS, KS}
{D(i) i = 40 to 52} = {AC, 2C, 3C, 4C, 5C, 6C, 7C, 8C, 9C, 10C, JC, QC, KC}

The model developed here assumes two players (A,B) and each are dealt five
cards from the deck of 52 cards. Altogether, ten cards are dealt, the first five to
player A and the next five to B.

Generate Random Hands to Players A and B

The routine below shows how to generate random hands of five cards each to
players A and B:

1. Using the random integer method of 52 unique integers, randomly generate
   n = 10 integers in sequence and label as: {E(j) j = 1 to 10}.
2. Use {E(j) j = 1 to 5} to select from the set D(i), the five cards for player A.
   Label as {C(k) k = 1 to 5}.
3. Use {E(j) j = 6 to 10} to select from the set D(i), the five cards for player B.
   Label as {C(k) k = 6 to 10}.

4. Return {C(k) k = 1 to 5} for player A, and {C(k) k = 6 to 10} for player B.

Example 7.8 The routine below shows how to deal five cards to players A and B from the deck of 52 cards, where the dealt cards are without replacement:

1. Using the Random Integer method with N = 52, generate n = 10 randomly sequenced integers, {E(1), ..., E(10)}. Say, {27, 5, 24, 16, 14, 32, 47, 31, 4, 25}.
2. For player A, the first five integers {27, 5, 24, 16, 14} yield {AS, 5H, JD, 3D, AD}.
3. For player B, the next five integers {32, 47, 31, 4, 25} yield {6S, 8C, 5S, 4H, QD}.
4. Return C(k) k = 1 to 5 = {AS, 5H, JD, 3D, AD} for player A, and C(k) k = 6 to 10 = {6S, 8C, 5S, 4H, QD} for player B.

Summary

This chapter concerns applications that are not from the common probability distributions, continuous or discrete. The applications are instructive since they show some popular deviations in generating random variates as is often needed in building computer simulation models. The applications presented are the Poisson process, constant Poisson process, batch arrivals, active redundancy, standby redundancy, random integers without replacement and poker.

Chapter 8
Output from Simulation Runs

Introduction

Computer simulation models are generally developed to study the performance of a system that is too complicated for analytical solutions. The usual goal of the analyst is to develop a computer simulation model that emulates the activities of the actual system as best as possible. Many of these models are from terminating and nonterminating systems.

A terminating system is when a defined starting event B and an ending event C are specified, and so, each run of the simulation model begins at B and ends at C. This could be a model of a car wash that opens each day at 6 a.m. and closes at 8 p. m. Each simulation run would randomly emulate the activities from B to C.

A nonterminating system is where there is no beginning or ending events to the system. The system often begins in a transient stage and eventually falls into either an equilibrium stage or a cyclical stage. This could be a study of a maintenance and repair shop that is always open. At the outset of the simulation model run, the system is empty and may take some time to enter either an equilibrium stage or a cyclical stage. This initial time period is called the transient stage.

A nonterminating system with transient and equilibrium stages might be a system where the inter-arrival flow of new customers to the shop is steadily coming from the same probability distribution. In the run of the simulation model, the system begins in the transient stage and thereafter the flow of activities continues in the equilibrium stage.

A nonterminating model with transient and cyclical stages could be a model of a system where the probability distribution of the inter-arrival flow of new customers varies by the hour of the day. The simulation run begins in a transient stage and passes to the cyclical stage thereafter.

In either system, while the analyst is developing the computer model, he/she includes code in the model to collect data of interest for later study. This output data is used subsequently to statistically analyze the performance of the system.

N.T. Thomopoulos, *Essentials of Monte Carlo Simulation: Statistical Methods for Building Simulation Models*, DOI 10.1007/978-1-4614-6022-0_8, © Springer Science+Business Media New York 2013

Terminating System

A terminating system is when at some point in time, or in events, the system comes to a natural end. Various portions of each run of the computer simulation model may be of interest to the analyst, and for each portion, a collection of output measures are saved for subsequent statistical analysis. To illustrate, an example is provided below.

Suppose an analyst is developing a computer simulation model of a large car wash system. Assume, the business opens at 8 a.m. and closes at 8 p.m; and at the beginning of the day, the car wash is empty. During the open hours, the customer arrival rate varies over the day. The end of the day is when the last car enters before closing time of 8 p.m. The analyst may be interested in the activities at different interval times of the day, perhaps from 8 to 12 a.m., 12 to 4 p.m. and 4 to 8 p.m. For convenience, the time periods are labeled as j ($j = 1$ to 3). Also suppose the measures of interest (denoted by index k), are collected for each time interval j. In the example, assume these are the following:

x_{1j} = number vehicle arrivals in j
x_{2j} = number vehicles serviced without waiting in j
x_{3j} = total vehicle wait time in j
x_{4j} = total system idle time in j
$p_{1j} = x_{2j}/x_{1j}$ = percent of vehicles that do not need to wait in j

Suppose further, the model is run n times, each with a different string of random variates, where i denotes the run number, whereby $i = 1$ to n. So the output data for the n runs of the simulation model would be the following:

$$x_{kji} \qquad k = 1 \text{ to } 4, \ j = 1 \text{ to } 3 \text{ and } i = 1 \text{ to } n$$

$$p_{1ji} \qquad k = 1, \ j = 1 \text{ to } 3 \text{ and } i = 1 \text{ to } n$$

As demonstrated, a large variety of output data could be collected for the simulation model and this is the type of data that will subsequently require some sort of statistical analysis to measure the performance of the system. Chapter 9 describes the typical statistical methods.

Nonterminating Transient Equilibrium Systems

The computer simulation model could be for a system that is nonterminating and evolves into an equilibrium state. The computer model would begin empty and flow through a transient stage prior to reaching the equilibrium stage. To illustrate, suppose the system under study is a mixed model assembly line where at the beginning of the simulation there are no units on the line. One by one, units are placed in station 1 and they move on up to station 2 and beyond, while new units are arriving to station 1. When the line is filled up in all stations, the transient staged ends and this is the start of the equilibrium stage. From that point on the system is in its equilibrium stage. In the example provided, this is fairly obvious; but in the general case, it is not obvious and a concern is how to determine when the transient stage ends.

Identifying the End of the Transient Stage

Suppose a variable of interest from the model is denoted as x_{ki}, the average of output index k in the i-th interval (batch) of events. The batches are run one after the other and the average for each batch is measured on every output index k. The difference from one batch to the previous is measured and the end of the transient stage is signaled when the differences begin to cycle around zero. The analyst seeks the event index i where thereafter, the average difference between two successive measures is sufficiently close to zero, i.e., $(x_{k,i+1} - x_{k,i}) \approx 0$. So, the transient stage, for output variable k, ends when such value of i is identified, say i = A, whereby the equilibrium stage follows. For convenience sake, the analyst would likely choose one value of A that defines the end of the transient stage for all K output variables.

Output Data

The output for each run begins after the transient period of A events has elapsed. When K is the number of output variables of interest, and n is the number of simulation runs during the equilibrium period, the output becomes the following:

$$x_{ki} \qquad k = 1 \text{ to K and } i = 1 \text{ to n.}$$

This is the output data that is subsequently analyzed by statistical methods.

Example 8.1 A one service facility with infinite capacity queuing system is under review with parameters τ_a = expected time between arrivals and τ_s = expected service time. The arrival rate is $\lambda = 1/\tau_a$ and the service rate is $\mu = 1/\tau_s$. The inter-arrival times and the service times are exponentially distributed. The utilization ratio is $\rho = \lambda/\mu = \tau_s/\tau_a$ where ρ must be less than one to attain an equilibrium system. This is a common queuing system listed in many textbooks. Two of the performance measures of this system is the expected time a unit is in the system, denoted as w, and another is the probability a new arrival is delayed in the queue before it enters the service facility, P_d. The analytical solution for this model is $w = 1/[(1 - \rho)\mu]$ and $P_d = \rho$.

A simulation model is developed for this system to demonstrate some of the concepts. Two set of parameters are shown, one is where $\rho = 0.10$ and another when $\rho = 0.50$. In both situations, $\mu = 1$. The simulation is run in batch sizes on $n = 500$ arrivals to measure the average time a unit spends in the system, w, and the probability a new arrival has to wait for service, P_d. The results are listed in the table below where arrivals reach 3,000. In the table, for each batch of $n = 500$ arrivals, w is the average time an arrival is in the system and P_d is the portion that are delayed in the queue before entering the service facility. The difference from one batch to another is measured as dw and dP_d. For example, at $\rho = 0.1$, $dw = (1.05 - 0.00) = 1.05$ for the first batch, $dw = (1.14 - 1.05) = 0.09$ for the second batch, and so forth. The end of the transient stage is signaled when the difference from one batch to another start to cycle around zero.

| | | $\rho = 0.10$ | | |
Runs	w	P_d	dw	dP_d
1–500	1.05	0.086	1.05	0.086
501–1,000	1.14	0.112	0.09	0.026
1,001–1,500	1.18	0.106	0.04	−0.006
1,501–2,000	1.05	0.116	−0.13	0.010
2,001–2,500	1.01	0.081	−0.04	−0.035
2,501–3,000	1.08	0.106	0.07	0.025

| | | $\rho = 0.50$ | | |
Runs	w	P_d	dw	dP_d
1–500	1.75	0.448	1.75	0.448
501–1,000	2.52	0.548	0.77	0.100
1,001–1,500	2.01	0.474	−0.51	−0.074
1,501–2,000	1.78	0.462	−0.23	−0.012
2,001–2,500	2.05	0.516	0.27	0.044
2,501–3,000	1.77	0.524	−0.28	0.008

At $\rho = 0.10$, note how dw and dP_d start to cycle around zero after 1,000 arrivals, and this signals the transient stage is ending at $n = 1,000$ arrivals. At $\rho = 0.50$, dw and dP_d both start to cycle around zero after 1,500 arrivals, indicating the transient stage ends at $n = 1,500$ arrivals, and the equilibrium stage begins.

Partitions and Buffers

Another way to collect the data for this system is described here. For all the events after the transent stage ends, the analyst could specify two parameters, N and M that will be used in partitions and buffers, respectively. Each partition is of length N and every buffer is of length M, where typically $N \geq M$. As the simulation model progresses, the partitions and buffers will follow in a leapfrog manner one after the other. For example, the progression of events is the following:

1 to N	Partition 1
N + 1 to N + M	Buffer 1
N + M + 1 to 2N + M	Partition 2
2N + M + 1 to 2N + 2M	Buffer 2
2N + 2M + 1 to 3N + 2M	Partition 3
. . .	

The run stops after n partitions.

Statistical measures will only include the data from the partitions, and not the buffers. The buffers are needed to allow the measures from one partition to the next to be far enough apart from each other where the status of one partition does not influence the status of the next. This way, the variable measures from the individual partitions are independent.

Suppose n is the number of partitions in the run and K is the number of output measures of interest that are collected for each partition. So, for output measure k of partition i, the output data saved would be x_{ki} where $k = 1$ to K and $i = 1$ to n.

Nonterminating Transient Cyclical Systems

The computer simulation model could be for a system that is nonterminating and progresses into a cyclical state. The computer model would begin empty and flow through a transient stage prior to reaching the cyclical stage. To illustrate, suppose the system under study is a car repair center where customers leave their vehicles for maintenance or repair. The shop is open 12 h a day and the vehicles remain in the system, even overnight, until the service is finished. The arrival rate varies by the time of day and as such the system follows a cyclical daily pattern. In the computer simulation model, after the transition stage ends, the status evolves into daily cycles.

Output Data

Upon identifying A, the number of events until the cyclical stage begins, the computer simulation model now runs with n different strings of random numbers. The output for each run begins after the transient period of A events have elapsed. When K is the number of output measures of interest and n is the number of runs, the output becomes the following:

$$x_{ki} \qquad k = 1 \text{ to } K \text{ and } i = 1 \text{ to n.}$$

This is the output data that is subsequently analyzed by statistical study.

Cyclical Partitions and Buffers

The output measures might be collected for each cycle and even at different intervals of the cycle. Suppose K is the number of variables of interest, J is the number of intervals in a cycle to measure, and n is the number of partitions. Hence, the data collected after n partitions is the following:

$$x_{kji} \quad \text{for } k = 1 \text{ to } K, \ j = 1 \text{ to } J \text{ and } i = 1 \text{ to } n$$

This would be the data for the subsequent statistical analysis.

To ensure the data from each cycle is independent, the concept of partitions and buffers could still be in place. To begin, the partitions would be the every second day cycle and the buffer would be the day in-between. So, the n partitions collected for the output would come from every second day of the simulation run. On some occasions, the analyst may select each third or fourth day to represent a partition.

Other Models

Some simulation models are not time or event related and have none of the associated traits described earlier as terminating, transient or equilibrium. Instead, the simulation model may be used to develop data that can be used in subsequent analysis. Several examples are provided below.

Forecasting Database

An analyst is testing a time series forecasting system that uses up to 24 months of historical demands and generates forecasts for the coming 12 months. The forecasts are of the horizontal, trend and seasonal type. The forecast system processes one part at a time and determines the forecast model to use and estimates the coefficients associated with the forecast model. The forecasts for each of the next 12 months are then generated.

To analyze the efficiency of the forecast system, the analyst wants to generate a database record for a series of parts to process through the forecast system. This requires the fields in the part record as: part number, description (optional), number months of history (1–24), and the 24 most recent history demands that are available (could be from horizontal, trend or seasonal). The simulation model would need to generate the data, one field at a time, in a way that is realistic and would allow the analyst to measure the efficiency of the forecasting system.

The simulation model could generate any number of part number records, 10, 100, 1,000, as needed. It would be good for each part record to have a comment field to identify the type of history data (horizontal, trend, seasonal) so the results could compare with the forecast model generated. The analyst could also occasionally insert an outlier demand in one of the history fields to see how competent the forecast system is to detect for outliers and adjust accordingly. The output record for each part might include the following:

Part, Comments, Number Months History, D1, D2, ….., D24.

Forecast and Replenish Database

In the event the system also has replenish capability, the forecast-replenish system would then compute the order size, safety stock, order point and order level. Depending on the on-hand and on-order data, the system would compute the replenish quantity needed, if any, for each part. The simulation part data now includes fields with the following: on-hand, on-order, cost per unit, multiple quantity, minimum buy quantity, lead time, and price break data when they pertain.

The simulation model has to coordinate this data for each part with the above forecast data generated to yield a realistic database record for each part. Comments should be included in a field to allow the analyst to compare the forecast-replenish system results with the data provided. When the on-hand and on-order are low, a replenish quantity should be called in the subsequent replenishment routine. When the on-hand is low and the on-order high, no replenish quantity is needed, and so forth. The additional data per part may include the following:

Part, Cost, Multiple Quantity, Min Quantity, Lead Time, On-Hand, On-Order, Price Break Data.

Example 8.2 An inventory manager is seeking guidance on how to set the forecasting parameter, α (alpha) that plays an important role in controlling the inventory. In particular, the horizontal single smoothing forecasting model is in use to generate the forecasts for the future months for many of the parts in the inventory. The particular values of α under consideration are: 0.05, 0.10, 0.20, 0.30, 0.40 and 0.50.

A simulation model is developed to randomly generate, for each part, 48 months of demand history that follows a horizontal demand pattern with a coefficient of variation (cov) set at 0.30. The demands are randomly generated using the normal distribution. The cov implies, the standard deviation of the demands is $\sigma = 0.30\mu$ where μ is the average demand per month. Essentially, $cov = \sigma/\mu$ where σ is the standard deviation of each monthly demand. So for each part in the simulation study, the demands generated are: $x_1, ….. ,x_{48}$.

The forecast model is run through the first 24 months of history and the forecast for the next 12 months of demands is generated. Starting with the month 24 forecast, the forecast errors are measured for each of the next 12 months, and the standard error of the forecast error is tabulated. From month 24 to month 36, the forecast model moves forward and the forecast errors are measured in the same way.

Altogether, 13 sets of forecasts are generated for each part, (one each for months 24–36), and the corresponding forecast errors are measured. Finally, the cov of the 1 month ahead forecast error is measured by $cov = s/\bar{x}$, where s is the standard deviation of 1 month ahead forecast error, and $\bar{x}$ is the average 1 month demand.

This process is followed for each of the six parameter values of α listed earlier, and also for 100 parts where the demand stream of 48 months are each generated with a different set of random numbers. The average cov from all of the 100 parts are listed in the table below for each of the α values under review. The results clearly show where the smaller value of the parameter α yields the best forecast results. The forecaster is cautiously aware that in the simulation model all the data are from a horizontal demand pattern and the accuracy results are for history patterns that are truly of that type.

α	Cov
0.05	0.297
0.10	0.305
0.20	0.315
0.30	0.318
0.40	0.325
0.50	0.344

Example 8.3 An inventory manager is concerned on the forecast accuracy for a part depending on the number of month's history (nmh) demand that is available to generate the forecast. The forecast manager is using the horizontal single smoothing model that generates the forecasts for the future months, for many of the parts in the inventory. The parameter for the forecast model is set as $\alpha = 0.10$.

A simulation model is developed to randomly generate 36 months of demand history that follows a horizontal demand pattern with a coefficient of variation (cov) set at 0.30. The demands are randomly varied using the normal distribution. The standard deviation of the demands is $\sigma = 0.30\mu$ where μ is the average demand per month. So for each part in the simulation study, the demands generated are: $x_1, \ldots, x_{36}$.

The forecast model is run through the first 24 months of history and the forecast for the next 12 months of demands is generated. The forecast errors are measured for each of the next 12 months from which the standard error of the forecast error is tabulated. Finally, the cov of the 1 month forecast error is measured by $cov = s/\bar{x}$, where s is the standard deviation of 1 month forecast error, and $\bar{x}$ is the average 1 month demand.

The table below lists the results from one such part by nmh. The table clearly shows how the cov improves as the nmh increases from 1 to 24.

Month	nmh	Cov
1	1	0.71
2	2	0.60
3	3	0.55
4	4	0.50
5	5	0.45
6	6	0.46
7	7	0.56
8	8	0.55
9	9	0.45
10	10	0.43
11	11	0.38
12	12	0.36
13	13	0.38
14	14	0.31
15	15	0.31
16	16	0.32
17	17	0.34
18	18	0.36
19	19	0.27
20	20	0.28
21	21	0.31
22	22	0.28
23	23	0.28
24	24	0.32

Example 8.4 An inventory manager is wondering whether to include logic in the forecasting model to detect and adjust the demand history for outlier demands prior to generating the forecasts. An outlier demand is where one (or more) months have a demand that is much larger (or smaller) than the normal flow of the other demands in the history. Of specific interest is the affect on the horizontal single smoothing model that generates the forecasts for the future months for many of the parts in the inventory. The parameter for the forecast model is set at $\alpha = 0.10$.

A simulation model is developed to randomly generate 36 months of demand history that follows a horizontal demand pattern with a coefficient of variation (cov) set at 0.30. The demands are randomly varied using the normal distribution. The standard deviation of the demands is $\sigma = 0.30\mu$ where μ is the average demand per month. So for each part in the simulation study, the demands generated are: $x_1, \ldots, x_{36}$. Three sets of demand history are generated for each part. The first set has no outliers. The next two sets have one outlier inserted in the demand history somewhere in months 1–24.

For each of the three sets (no outlier, one outlier, one outlier), the forecast model is run through the first 24 months of history and the forecast for the next 12 months of demands is generated. The forecast errors are measured for each of the next 12 months from which the standard error of the forecast error is tabulated. Finally, the cov of the 1 month forecast error is measured by $cov = s/\bar{x}$, where s is the standard deviation of 1 month forecast error, and $\bar{x}$ is the average 1 month demand.

The table below lists the results from each of the three sets of demand history from the same part. The table shows how the cov increases tremendously when the part has one outlier in the demand history. The forecast manager can now clearly see how important it is to include logic in the forecasting routine to detect for outlier demands and adjust accordingly prior to forecasting.

Part	Demand history	Cov
1	No outlier	0.343
2	One outlier	0.611
3	One outlier	0.476

Example 8.5 An inventory manager is seeking guidance on various decisions concerning the seasonal forecasting model. One of the decisions concerns the three parameters to the model: α (for the average), β (for the trend) and γ (for the seasonal pattern). The particular values of α under consideration are: 0.05, 0.10, 0.20, 0.30, 0.40 and 0.50. The values for β and γ both are (0.10 and 0.20).

A simulation model is developed to randomly generate, for each part, 48 months of demand history that follows a seasonal demand pattern with a coefficient of variation (cov) set at 0.30. The demands are randomly generated using the normal distribution. So for each part in the simulation study, the demands generated are: $x_1, \ldots, x_{48}$.

The forecast model is run through the first 24 months of history and the forecast for the next 12 months of demands is generated. Starting with the month 24 forecast, the forecast errors are measured for each of the next 12 months, and the standard error of the forecast error is tabulated. From month 24 to month 36, the forecast model moves forward and the forecast errors are measured in the same way. Altogether, 13 sets of forecasts are computed for each part, (one each for months 24–36), and the corresponding forecast errors are measured. Finally, the cov of the 1 month ahead forecast error is measured by $cov = s/\bar{x}$, where s is the standard deviation of 1 month ahead forecast error and $\bar{x}$ is the average 1 month demand.

This process is followed for each combination of the parameter values of α, β, γ listed earlier, and also for 100 parts where the demand stream of 48 months are each generated with a different set of random numbers. The average cov from all of the 100 parts are listed in the table below for each parameter (α, β, γ) combination. The results show where the smaller value of the parameters yields the best results. The forecaster is cautiously aware that all the data are from a seasonal demand pattern and the accuracy results are for history patterns that are truly of that type.

Cov when the demand pattern is seasonal and the seasonal forecast model is run with the following parameter values:

α	$\beta = 0.10$ $\gamma = 0.10$ cov	$\beta = 0.10$ $\gamma = 0.20$ cov	$\beta = 0.20$ $\gamma = 0.10$ cov	$\beta = 0.20$ $\gamma = 0.20$ cov
0.05	0.306	0.316	0.307	0.312
0.10	0.314	0.317	0.315	0.329
0.20	0.318	0.325	0.320	0.333
0.30	0.337	0.338	0.347	0.357
0.40	0.341	0.351	0.347	0.357
0.50	0.357	0.366	0.369	0.383

Example 8.6 An inventory manager is seeking guidance on various decisions concerning the seasonal demand pattern. One of the decisions concerns the forecast model to apply: horizontal, trend or seasonal. The horizontal model uses the parameter α; the trend model uses parameters α, β; and the seasonal model uses the parameters: α (for the average), β (for the trend) and γ (for the seasonal pattern). The particular values of α under consideration are: 0.05, 0.10, 0.20, 0.30, 0.40 and 0.50. The values for β and γ are (0.10 and 0.10).

A simulation model is developed to randomly generate, for each part, 48 months of demand history that follows a seasonal demand pattern with a coefficient of variation (cov) set at 0.30. The demands are randomly generated using the normal distribution. So for each part in the simulation study, the demands generated are: $x_1, \ldots, x_{48}$. The forecast model is run through the first 24 months of history and the forecast for the next 12 months of demands is generated. Starting with the month 24 forecast, the forecast errors are measured for each of the next 12 months, and the standard error of the forecast error is tabulated. From month 24 to month 36, the forecast model moves forward and the forecast errors are measured in the same way. Altogether, 13 sets of forecasts are generated for each part, (one each for months 24–36), and the corresponding forecast errors are measured. Finally, the cov of the 1 month ahead forecast error is measured by cov $= s/\bar{x}$, where s is the standard deviation of 1 month ahead forecast error, and $\bar{x}$ is the average 1 month demand.

This process is followed for each combination of the parameter values of α, β, γ listed earlier, and also for 100 parts where the demand stream of 48 months are each generated with a different set of random numbers. Each of the three forecast models are run with the same data. The average cov from all of the 100 parts are listed in the table below for each forecast model and of every α, β, γ combination. The results clearly show that the best forecasts for the data from a seasonal demand pattern are those that are generated from the seasonal forecast model. The better forecasts are also those with the smaller parameter values.

Cov when the demand pattern is seasonal and the forecast models (horizontal, trend, seasonal) are run with the following parameter values:

| | -------------Forecast model------------- | | |
	horizontal	trend $\beta = 0.10$	seasonal $\beta = 0.10$ $\gamma = 0.10$
α	cov	cov	cov
0.05	0.443	0.404	0.306
0.10	0.429	0.419	0.314
0.20	0.424	0.435	0.325
0.30	0.423	0.442	0.337
0.40	0.409	0.428	0.341
0.50	0.400	0.418	0.357

Summary

Computer simulation models are mainly developed to emulate actual systems that are too complex to analyze mathematically. The systems often fall into the terminating or the nonterminating type. The chapter describes how output data is collected for either type of system. Terminating systems have a defined beginning and ending event, and nonterminating systems include a combination of transient, equilibrium and cyclical stages. The output data from these systems are needed subsequently to statistically analyze the performance of the system that is under study. Another simulation model presented is one that creates the database to be used as test data for software applications like forecasting and inventory replenishments.

Chapter 9
Analysis of Output Data

Introduction

This chapter is a quick review on some of the common statistical tests that are useful in analyzing the output data from runs of a computer simulation model. This pertains when each run of the model yields a group of k unique output measures that are of interest to the analyst. When the model is run n times, each with a different string of continuous uniform u ~ U(0,1) random variates, the output data is generated independently from run to run, and therefore the data can be analyzed using ordinary statistical methods. See, for example, Hines et al. (2003) for a full description on statistical methods. Some of the output data may be of the variable type and some may be of the proportion type. The appropriate statistical method for each type of data is applied as needed. This includes, measuring the average value and computing the confidence interval of the true mean. Oftentimes, the simulation model is run with one or more control input variables in a 'what if' manner. The output data between the two or more settings of the control variables can be compared using appropriate statistical tools. This includes testing for significant difference between two means, between two proportions, and between k or more means.

Example 9.1 A maintenance and repair shop for cars is open Monday through Saturday from 8 a.m. till 6 p.m. The cars needing service arrive during the day with an average arrival rate via the Poisson distribution. The service times vary via a gamma distribution with a location parameter to signify the minimum time of service. A simulation model is developed to emulate the daily activities and the model collects a series of measures of interest to the analyst. The number of independent runs of the model is n. Some of the measures collected in each run of the model are listed below:

$n0$ = number of bays (this is a control parameter)
$n1$ = number of vehicles that arrive for service.
$n2$ = number of parts needed to complete the service.
$n3$ = number of vehicles serviced without a delay.

N.T. Thomopoulos, *Essentials of Monte Carlo Simulation: Statistical Methods for Building Simulation Models*, DOI 10.1007/978-1-4614-6022-0_9,
© Springer Science+Business Media New York 2013

n4 = number of needed parts that are available in the stock room.
n5 = the number of vehicles that wait in queue over 60 min.
S1 = sum wait time for delayed vehicles.
S2 = sum idle time for a repair bay over the day.

Variable Type Data

The variable type data for the individual daily run are the following:

x1 = S1/n1 = average wait time per vehicle.
x2 = S2/(10 × n0) = average idle time per hour per bay.
x3 = n3/n1 = service level for the vehicles.
x4 = n4/n2 = service level for the parts.

Proportion Type Data

When n replications of the simulation model are run, and n5 is the number of days
when one or more vehicles wait at least 60 min for service, then,
    p = n5/n is the proportion of days when a vehicle waits 60 or more minutes.

Analysis of Variable Type Data

Variable type data is like time to complete a service on an item in repair, strength
of a steel beam, elasticity from rubber compound, labor cost to service a vehicle,
and so forth. Upon completion of n replications of a simulation model, the analyst
may be interested in determining the point estimate on the variable measured in
the model and also the corresponding confidence interval. The analyst may
further inquire on how many replications are needed to gain the precision desired
in the estimate.

Consider the output measure, x, from a run of a simulation model. The variable x
could be the average of a large number of events that take place in the model. In the
service department of the car dealership model, x could be the average time a
customer waits for service on the vehicle. In the typical situation, the distribution
and the mean and variance of x are unknown. When n replications of the simulation
are run, with the same initial conditions and with different streams of continuous
uniform $u \sim U(0,1)$ random variates, then the output data, $x_1, \ldots, x_n$, for each
random variable can be treated as independent observations of the random variable.

Some of the more common statistical tools for analyzing the output data are presented below. Each of the tools are valid as long as the n sample observations, $x_1, \ldots, x_n$, are statistically independent.

Sample Mean and Variance

Using the n output results, $x_1, \ldots, x_n$, the sample mean and variance of x are computed as below:

$$\bar{x} = \sum_{i=1}^{n} x_i/n$$

$$s^2 = \sum_{i=1}^{n} (x_i - \bar{x})^2/(n-1)$$

The standard deviation of x is $s = \sqrt{s^2}$, and the standard error of the mean, $s_{\bar{x}}$, is obtained by,

$$s_{\bar{x}} = s/\sqrt{n}$$

The associated degrees of freedom is $(n-1)$.

So now, $\bar{x}$ is an estimate of the true mean, μ, and s is an estimate of the true standard deviation, σ. Note the true values of the parameters (μ and σ) are unknown. Also not known is the distribution of x.

Confidence Interval of μ when x is Normal

In the event, x is normally distributed, the $(1 - \alpha)$ confidence interval of μ becomes,

$$L \leq \mu \leq U$$

where, the upper and lower confidence limits (U, L) are, obtained from

$$U = \bar{x} + t_{\alpha/2}\, s_{\bar{x}}$$

$$L = \bar{x} - t_{\alpha/2}\, s_{\bar{x}}$$

Note, $t_{\alpha/2}$, has $(n - 1)$ degrees of freedom and is the $\alpha/2$ upper-tail percentage-point of the student's t distribution where $P(t > t_{\alpha/2}) = \alpha/2$. Hence,

$$P(L \leq \mu \leq U) = 1 - \alpha.$$

In the event n is large, say n > 30, the standard normal variable, z, can be used in place of the student's variable, t. In this event, the confidence limits become,

$$U = \bar{x} + z_{\alpha/2}\, s_{\bar{x}}$$

$$L = \bar{x} - z_{\alpha/2}\, s_{\bar{x}}$$

where $z_{\alpha/2}$ is the z value from $N(0,1)$ that gives $P(z > z_{\alpha/2}) = \alpha/2$.

Approximate Confidence Interval of μ when x is Not Normal

In the event, x is not normally distributed, and the sample size is small, an approximate $(1 - \alpha)$ confidence interval of μ is computed in the same way. That is,

$$L \leq \mu \leq U$$

where, the upper and lower confidence limits (U, L) are, obtained from

$$U = \bar{x} + t_{\alpha/2}\, s_{\bar{x}}$$

$$L = \bar{x} - t_{\alpha/2}\, s_{\bar{x}}$$

The term $t_{\alpha/2}$ is the upper-tail percentage-point from the student's t distribution with degrees of freedom $(n - 1)$, where $P(t > t_{\alpha/2}) = \alpha/2$. But since, x is not normal, the exact probability of the interval is not truly known and is approximated as,

$$P(L \leq \mu \leq U) \approx 1 - \alpha.$$

Central Limit Theorem

When n increases, the Central Limit Theorem applies, and the distribution shape of the sample mean approaches a normal distribution. Hence, the standard normal variable, z, replaces the student's variable, t and the confidence limits become,

$$U = \bar{x} + z_{\alpha/2}\, s_{\bar{x}}$$

$$L = \bar{x} - z_{\alpha/2}\, s_{\bar{x}}$$

where $z_{\alpha/2}$ is the z value that gives $P(z > z_{\alpha/2}) = \alpha/2$. The confidence interval on μ becomes,

$$P(L \le \mu \le U) = 1 - \alpha.$$

Example 9.2 Suppose a simulation run yields a variable, x with each trial run of the simulation. Assume further, the simulation is run with $n = 10$ repetitions, and each repetition begins with the same initial values and terminates over the same length of events. The only difference is the stream of random variates used in each of the simulation runs. So, as much as possible, there are now n output results, x_1, ..., x_n, that are from the same distribution and are independently generated. Assume further, the sample mean and variance from the ten samples are the following:

$$\bar{x} = 20.0$$

and

$$s^2 = 25.0,$$

respectively. The standard deviation becomes $s = 5$, and the standard error of the mean is,

$$s_{\bar{x}} = s/\sqrt{n} = 1.58.$$

Also, the degrees of freedom is $(10 - 1) = 9$ for this data.

Because the true distribution of x is not known, an approximate confidence interval is computed. The student's t variable needed for a 95 % confidence is $t_{0.025}$. Since the degrees of freedom is 9, the search of the student t distribution yields, $t_{0.025} = 2.262$. The approximate upper and lower 95 % confidence limits on μ can now be computed and become,

$$U = 20 + 2.262 \times 1.58 = 23.57$$

$$L = 20 - 2.262 \times 1.58 = 16.43$$

The corresponding approximate confidence interval is $(16.43 \le \mu \le 23.57)$, where

$$P(16.43 \le \mu \le 23.57) \approx 0.95$$

When Need More Accuracy

Suppose the analyst wants the length of the $(1 - \alpha)$ confidence interval, currently $(U - L)$, to shrink to $2E$, and all else remains the same. The number of repetitions to achieve this goal is obtained from the relation below,

$$n = [(s \times t_{\alpha/2})/E]^2$$

where s is the sample standard deviation, t is the student's t value and E is the precision sought. Note the student's t value must be coordinated with degree of freedom $(n - 1)$. The problem with the above relation is that the t value cannot be inserted in the formula until the sample size n is known.

A way to approximate the formula is to use the normal z value instead of the student's t value in the above formula. Replacing t with z yields,

$$n = [(s \times z_{\alpha/2})/E]^2$$

The above estimate of n will be less or equal to the counterpart value of n when the student's t is used. As n gets larger, $(n > 30)$, the difference between using the t and z value is minor.

Example 9.3 Consider the approximate 95 % confidence interval that is shown in Example 9.2. Suppose the analyst wants the length of the 95 confidence interval to shrink from the current $(U - L) = (23.57 - 16.43) = 7.14$ to $(U - L) = 4.0$. The number of repetitions to achieve this goal is obtained from the relation below,

$$n = [(s \times z_{\alpha/2})/E]^2 = [(5 \times 1.96)/2]^2 = 24.01$$

So, in this example, n = 24 repetitions are needed. Because, n = 10 repetitions have already been run, 14 new repetitions are needed.

Analysis of Proportion Type Data

Proportion type data is like the portion of trials an event occurs. Examples are: the portion of units that have defects; the portion of customers who use a credit card for a purchase; the portion of customers in a gas station that use premium gas; the portion of police calls that have to wait more than 10 min for service, and so forth. Upon completion of n replications of a simulation model, the analyst may be interested in determining the point estimate of the proportion measured in the model, and also on the corresponding confidence interval. The analyst may also inquire how many replications are needed to gain more precision.

Proportion Estimate and Its Variance

Using the n output results, let w represent the number from the n replications where a specified event occurs. The estimated proportion is obtained by,

$$\hat{p} = w/n$$

The corresponding variance is

$$s\hat{p}^2 = \hat{p}(1 - \hat{p})/n$$

and the standard error of $\hat{p}$ becomes,

$$s\hat{p} = \sqrt{\hat{p}(1 - \hat{p})/n}$$

Confidence Interval of p

The distribution on the estimate of p is approximated by the normal distribution when $np \geq 5$ at $p \leq 0.50$, or when $n(1 - p) \geq 5$ at $p \geq 0.50$. So when the number of replications, n, is sufficiently large, the $(1 - \alpha)$ confidence interval on the true proportion, p, can be computed in the following way. Let U and L be the upper and lower confidence limits, respectively, whereby,

$$L \leq p \leq U$$

The confidence limits are computed as follows,

$$U = \hat{p} + z_{\alpha/2}\, s\hat{p}$$

$$L = \hat{p} - z_{\alpha/2}\, s\hat{p}$$

The probability on the confidence interval is,

$$P(L \leq p \leq U) = (1 - \alpha).$$

Example 9.4 Suppose a terminating simulation model is run with $n = 60$ replications, and an event A occurs on $w = 6$ of the replications. The analyst wants to estimate the portion of times the event will occur. The estimate of the portion of times event A occurs is

$$\hat{p} = 6/60 = 0.10.$$

The variance of the estimate is the following.

$$s\hat{p}^2 = \frac{0.10\,(1 - 0.10)}{60} = 0.0015$$

and the standard error of p becomes,

$$s\hat{p} = \sqrt{0.0015} = 0.039$$

Since the estimate of p ($\hat{p} = 0.10$) is less than 0.50 and $n(1 - p) = 6.0$ is larger than 5, the normal approximation can be applied and the confidence limit is computed as below. Recall, $z_{0.025} = 1.96$.
   The 95 % confidence limits on p becomes,

$$U = 0.10 + 1.96 \times 0.039 = 0.176$$

$$L = 0.10 - 1.96 \times 0.039 = 0.024$$

whereby, the 95 % confidence interval is

$$(0.024 \le p \le 0.176)$$

and the associated probability is

$$P(0.024 \le p \le 0.176) = 0.95.$$

When Need More Accuracy

In the event the analyst desires more accuracy on the estimate of the proportion, p, the number of repetitions needs to increase. The following formula computes the size of n for the accuracy desired. Suppose $(1 - \alpha)$ is fixed and the tolerance $E = 0.5(U - L)$ desired is specified. So the number of repetitions becomes,

$$n = p(1 - p)[z_{\alpha/2}/E]^2$$

If an estimate on the proportion p is not available, then set $p = 0.5$, and the number of repetitions becomes

$$n = (0.25)[z_{\alpha/2}/E]^2$$

Example 9.5 Consider Example 9.4 again, and assume the analyst needs more accuracy in the estimate of p. He/she wants to lower the 95 % tolerance from 0.5 $(0.176 - 0.24) = 0.076$ to $E = 0.050$. The question is how many more repetitions are needed to accomplish. The above formula is used assuming $p = 0.10$ and letting $(1 - \alpha) = 0.95$. Hence, $\alpha = 0.05$ and $z_{\alpha/2} = 1.96$. The number of repetitions needed becomes,

$$n = (0.1)(0.9)[1.96/0.05]^2 = 138.3.$$

Since, 60 repetitions have already been run, 79 more are needed.

Example 9.6 A machine shop has an order for ten units (No $= 10$) of a product that require processing on two machines, M_1 and M_2. One component is produced on M_1 and another on M_2. The two components are combined to yield the final product. The machine processing is expensive and difficult where defective components can occur on each machine. The management of the shop wants to determine in advance how many units to start, Ns, at the outset to be 95 % certain the number of good units, Ng, of the final product is equal or larger than the required units of No $= 10$. In essence, they want to determine Ns where $P(Ng \geq No) \geq 0.95$.

When M_1 launches Ns units at the start, the raw material is gathered and the units are processed one after the other. The number of good units from M_1 is denoted as g_1, whereby the number of defective units is $d_1 = Ns - g_1$. A defective unit from processing on M_1 can occur in two ways, (1) when the raw material is defective, and (2) when the processed unit fails a strength test. The probability of a defective raw material is $P(d) = 0.03$. The strength of the material, S_1, is equally likely to fall as $10 \leq S_1 \leq 20$. The force, F_1, is distributed as an exponential distribution with expected value $E(F_1) = 5.0$. If $F_1 > S_1$, the unit is defective since the strength is not adequate.

When M_2 launches Ns units at the start, the raw material is gathered and the units are processed one after the other. The number of good units from M_2 is denoted as g_2, whereby the number of defective units is $d_2 = Ns - g_2$. A defective unit from processing on M_2 can occur in two ways, (1) when the raw material is defective, and (2) when the processed unit fails a strength test. The probability of a defective raw material is $P(d) = 0.05$. The strength of the material, S_2, is equally likely to fall as $15 \leq S_2 \leq 24$. The force, F_2, is distributed as an exponential distribution with expected value $E(F_2) = 6.0$. If $F_2 > S_2$, the unit is defective since the strength is not adequate.

The good units from M_1 and M_2 are combined to yield the final product. The number of good units in the final product is $Ng = Min(g_1, g_2)$. Recall the goal is to begin Ns units on both machines so that the probability of $Ng \geq No$ is 0.95 or better.

A simulation model is developed to guide the management on the size of Ns to apply.

The results are listed below where the number of units to start, Ns, ranges from 10 to 15. For each attempt at Ns, the simulation is run 1,000 times to estimate the probability that the number of good units will be ten or larger. At Ns $= 10$, for example, the probability is 0.156 indicating that on 156 of the 1,000 trials, Ng was ten or larger, and this is far short of the goal. Note at Ns $= 13$, the probability reaches 0.952 and this is the minimum value of Ns to achieve the specified goal of the management.

No	Ns	$P(Ng \geq No)$
10	10	0.156
10	11	0.537
10	12	0.850
10	13	0.952
10	14	0.988
10	15	0.999

Example 9.7 In Example 9.6, from 1,000 trials with control variable Ns = 13 as the number of units to start production, the number of runs producing ten or more good units becomes 952. This is a proportion situation where the estimate of the proportion of good units, p, becomes $\hat{p} = 952/1000 = 0.952$. The associated standard deviation of this estimate is $s_p = \sqrt{(.952 \times .048)/1000} = .0067$. Hence, the 90 % confidence limits of p are,

$$L = 0.952 - 1.645 \times .0067 = 0.941$$

$$U = 0.952 + 1.645 \times .0067 = 0.963$$

Finally, the 90 % confidence interval is

$$P(0.941 \leq p \leq 0.963) = 0.90.$$

Based on the confidence interval, since L = 0.941 is less than 0.95, the management should be aware, with 90 % confidence, the proportion of good units could be lower than the goal of 0.95, when Ns = 13. Instead of settling on Ns = 14 units to start the production process, the analyst could consider taking more samples to gain a higher precision on the estimate of p at Ns = 13.

Comparing Two Options

Sometimes the analyst is seeking the better solution when two or more control options are in consideration. An example could be a mixed model assembly line where k different models are produced on the line and the analyst is seeking the best way to send the different models down the line for the day. One way is to send each model down the line in batches, and another way is to send them down the line in a random order. The assembly time for each model is known by station. A simulation model is developed and is run for a day's activity and the following type of measures are tallied. Idle time is recorded when an operator in a station must wait for the next unit in sequence before he/she can begin its work. Congestion is when the operator has extended work on a unit and is forced to continue working on the unit even when the next unit arrives at the station on a moving conveyer. A goal is to minimize the idle time and congestion time over the day's activities.

S1 = sum of idle time across all stations and models for the day.
S2 = sum of congestion time across all stations and models for the day.
N1 = number of units assembled for the day.
N2 = number of stations on the line.
N3 = number of units with congestion over the day.

The computations for the day are the following:

$x_1 = S1/(N1 \times N2)$ = average idle time per station per unit assembled.

$x_2 = S2 /(N1 \times N2)$ = average congestion time per station per unit assembled.

$p_1 = N3/N1$ = proportion of units with congestion.

Note, the measures x_1 and x_2 are variable type data, and p_1 is proportion type data.

Comparing Two Means when Variable Type Data

Suppose a terminating simulation model has two different options (1 and 2) and the analyst wants to compare the one against the other to see which is more preferable. For option 1, n_1 repetitions of simulation runs are taken and the output yields a sample mean $\bar{x}_1$ and variance s_1^2. For option 2, n_2 simulation runs are generated and the results give $\bar{x}_2$ and s_2^2. Typically, the number of simulation runs are the same, whereby $n_1 = n_2$. The true mean and variance of options 1 and 2 are not known and are estimated by the sample simulation runs.

Comparing x_1 and x_2

The estimate of the difference between the true means is measured by, $\bar{x}_1 - \bar{x}_2$. Denoting the true means of options 1 and 2 by μ_1 and μ_2, respectively, the difference between the sample means is an estimate of the difference between the true means. The expected value of the difference yields the following,

$$E(\bar{x}_1 - \bar{x}_2) = (\mu_1 - \mu_2)$$

Confidence Interval of $(\mu_1 - \mu_2)$ when Normal Distribution

When both variables x_1 and x_2 are normally distributed, it is possible to compute a $(1 - \alpha)$ confidence interval on the difference between the two means, $(\mu_1 - \mu_2)$. The result will yield upper and lower confidence limits, (U, L) where,

$$L \le (\mu_1 - \mu_2) \le U$$

and

$$P[L \le (\mu_1 - \mu_2) \le U] = (1 - \alpha)$$

The lower and upper limits are computed in the following way,

$$U = (\bar{x}_1 - \bar{x}_2) + t_{\alpha/2}\, s_{(\bar{x}_1 - \bar{x}_2)}$$

$$L = (\bar{x}_1 - \bar{x}_2) - t_{\alpha/2} s_{(\bar{x}_1 - \bar{x}_2)}$$

Where

$s_{(\bar{x}_1 - \bar{x}_2)}$ = standard error of the difference between the two means,

$t_{\alpha/2}$ = the student's t value with degrees of freedom df.

The way to compute the above standard error and degrees of freedom will be given subsequently.

Significant Test

The significance of the difference between the two options can be noted by use of the confidence interval by observing the range of values from confidence limits L to U. In the event the interval (L to U) passes through zero, the means of the two options are not significantly different with $(1 - \alpha)$ confidence level. When the interval is always positive, the mean of option 1 is significantly higher than the mean of option 2. On the other hand, if the interval is always negative, the mean of option 1 is significantly smaller than the mean of option 2.

When $\sigma_1 = \sigma_2$

When the true standard deviations of the two options are assumed the same, the way to measure the standard error of the difference between the two means is shown below.

First, the two sample variances are combined to estimate the common variance. The is called the *pooled* estimate of the variance and is computed as follows,

$$s^2 = \left[(n_1 - 1)s_1{}^2 + (n_2 - 1)s_2{}^2\right] / [n_1 + n_2 - 2]$$

Second, the standard error on the difference between the two means becomes,

$$s_{(\bar{x}_1 - \bar{x}_2)} = s\sqrt{\frac{1}{n_1} + \frac{1}{n_2}}$$

and the corresponding degrees of freedom is,

$$df = (n_1 + n_2 - 2).$$

When $\sigma_1 \neq \sigma_2$

When the true standard deviations of the two options are not assumed the same, the way to measure the standard error of the difference between the two means is below:

$$s_{(\bar{x}_1 - \bar{x}_2)} = \sqrt{\frac{s_1^2}{n_1} + \frac{s_2^2}{n_2}}$$

The corresponding degrees of freedom, df, is computed in the following way. First, the variance error of the mean for options 1 and 2 are measured as below,

$$s\bar{x}_1^2 = s_1^2 / n_1$$

$$s\bar{x}_2^2 = s_2^2 / n_1$$

The degrees of freedom becomes,

$$df = [s\bar{x}_1^2 + s\bar{x}_2^2]^2 / [s\bar{x}_1^4 / (n_1 + 1) + s\bar{x}_2^4 / (n_2 + 1)] - 2$$

Approximate Confidence Interval of $(\mu_1 - \mu_2)$ when Not Normal

When one or both variables x_1 and x_2 are not normally distributed, it is possible to compute an approximate $(1 - \alpha)$ confidence interval on the difference between the two means, $(\mu_1 - \mu_2)$ in the same way as shown above when the normal distribution applies. The result will yield approximate upper and lower confidence limits, (U, L) where,

$$L \le (\mu_1 - \mu_2) \le U$$

and

$$P[L \le (\mu_1 - \mu_2) \le U] \approx (1 - \alpha)$$

The upper and lower limits are computed in the following way,

$$U = (\bar{x}_1 - \bar{x}_2) + t_{\alpha/2} \, s_{(\bar{x}_1 - \bar{x}_2)}$$

$$L = (\bar{x}_1 - \bar{x}_2) - t_{\alpha/2} s_{(\bar{x}_1 - \bar{x}_2)}$$

As Degrees of Freedom Increases

Via the central limit theorem, as the degrees of freedom increases, the shape of the distribution of $(\bar{x}_1 - \bar{x}_2)$ increasingly resembles a normal distribution, and eventually the approximation term in the confidence interval is dropped.

Example 9.8 Suppose a terminating simulation model has two options, (1,2), and ten simulation runs of each are taken. The goal is to compare the difference between the means of each option. Assume the sample mean and variance of the two options are computed and the results are listed below.

Option 1: $n_1 = 10$, $\bar{x}_1 = 50$, and $s_1^2 = 36$.

Option 2: $n_2 = 10$, $\bar{x}_2 = 46$, and $s_1^2 = 27$.

The analyst assumes the variables x_1 and x_2 are sufficiently close to a normal distribution, and also the variances of the two options are equal. Hence, the standard error of the difference between the two means is computed as follows.

First, the pooled estimate of the variance is calculated as below.

$$s^2 = [(10 - 1)36 + (10 - 1)27]/[10 + 10 - 2]$$
$$= 31.5$$

The pooled standard deviation becomes,

$$s = \sqrt{31.5} = 5.61$$

Second, the standard error on the difference between the means is calculated as,

$$s_{(\bar{x}_1 - \bar{x}_2)} = 5.61\sqrt{\frac{1}{10} + \frac{1}{10}}$$

$$= 2.51$$

The associated degrees of freedom is df $= (n_1 + n_2 - 2) = 18$.

To compute the $(1 - \alpha) = 0.95$ confidence interval, the lower and upper confidence limits are found in the following way.

The student's t variable for $\alpha/2 = 0.025$ and for 18 degrees of freedom is searched on the student's t table to find,

$$t_{0.025} = 2.101$$

Hence, the upper and lower confidence limits are computed as below,

$$U = (50 - 46) + 2.101 \times 2.51 = 5.27$$

$$L = (50 - 46) - 2.101 \times 2.51 = -1.27$$

The 0.95 confidence interval is,

$$-1.27 \le (\mu_1 - \mu_2) \le 5.27.$$

Note, the range of the confidence interval passes across zero. Hence, with the sample sizes taken so far, there is no evidence of a significant difference between the means of the two options at the $(1 - \alpha) = 95\,\%$ confidence level.

Comparing the Proportions Between Two Options

Suppose a terminating simulation model has two different options (1 and 2) and the analyst wants to compare the one against the other to see which is more preferable. For option 1, n_1 repetitions of simulation runs are taken and x_1 of them have an attribute of interest. The proportion of repetitions that have the attribute is measured by $\hat{p}_1 = x_1/n_1$. For option 2, n_2 repetitions are taken, x_2 have the attribute and the proportion becomes $\hat{p}_2 = x_2/n_2$. Typically, the number of simulation runs are the same, whereby $n_1 = n_2$. The true proportions of options 1 and 2 are not known and are estimated by the sample simulation runs.

Comparing p_1 and p_2

The estimate of the difference between the two true proportions p_1 and p_2 is measured by the difference of their estimates $\hat{p}_1$ and $\hat{p}_2$. The expected value of the difference yields the following,

$$E(\hat{p}_1 - \hat{p}_2) = (p_1 - p_2)$$

and the estimate of the difference in the two proportions is $\hat{p}_1 - \hat{p}_2$.

Confidence Interval of ($p_1 - p_2$)

When a sufficient number of repetitions (n_1, n_2) are taken, the normal distribution applies to the shape of the difference between $\hat{p}_1$ and $\hat{p}_2$, and it is possible to compute a $(1 - \alpha)$ confidence interval on the difference between the two proportions, ($p_1 - p_2$). The result will yield upper and lower confidence limits, (U, L) where,

$$L \leq (p_1 - p_2) \leq U$$

and

$$P[L \leq (p_1 - p_2) \leq U] = (1 - \alpha)$$

The lower and upper limits are computed in the following way,

$$L = (\hat{p}_1 - \hat{p}_2) - z_{\alpha/2} s_{p1-p2}$$

$$U = (\hat{p}_1 - \hat{p}_2) + z_{\alpha/2} s_{p1-p2}$$

where

$s_{p1 - p2} = $ standard error of the difference between the two proportions,
$z_{\alpha/2} = $ the standard normal variable where $P(z > z_{\alpha/2}) = \alpha/2$.

The standard error $s_{p1 - p2}$ is calculated as follows,

$$s_{p1}^2 = \hat{p}_1(1 - \hat{p}_1)/n_1$$

$$s_{p2}^2 = \hat{p}_2(1 - \hat{p}_2)/n_2$$

$$s_{p1-p2} = \left[s_{p1}^2 + s_{p2}^2\right]^{0.5}$$

Significant Test

The significance of the difference between the two options can be noted by use of the confidence interval by observing the range of values from confidence limits L to U. In the event the interval (L to U) passes through zero, the proportions of the two options are not significantly different with $(1 - \alpha)$ confidence level. When the interval is always positive, the proportion of option 1 is significantly higher than the proportion of option 2. On the other hand, if the interval is always negative, the proportion of option 1 is significantly smaller than the proportion of option 2.

Example 9.9 Suppose a terminating simulation model is run with two options to determine which is preferable with respected to an event, A. Option 1 is run with $n_1 = 200$ repetitions and event A occurs on $x_1 = 28$ occasions. Hence, $\hat{p}_1 = x_1/n_1 = 0.14$ is the portion of times that event A has occurred. Option 2 is run with $n_2 = 200$ repetitions and event A occurs on $x_2 = 44$ occasions, whereas, $\hat{p}_2 = x_2/n_2 = 0.22$. The analyst wants to determine the 95 % confidence interval on the difference between the two proportions, $(p_1 - p_2)$.

The point estimate on the true difference between the two proportions is,

$$(\hat{p}_1 - \hat{p}_2) = (0.14 - 0.22) = -0.08.$$

The variance of each of the proportions is,

$$s_{p_1^2} = (0.14)(1 - 0.14)/200 = 0.00060.$$

$$s_{p_2^2} = (0.22)(1 - 0.22)/200 = 0.00086.$$

The standard error between the two proportions is now computed as below,

$$s_{p1-p2} = \sqrt{0.00060 + 0.00086}$$
$$= 0.0382$$

To compute the confidence interval with $(1 - \alpha) = 0.95$, we need the standard normal value of $z_{0.025} = 1.96$. The confidence limits become,

$$U = (0.14 - 0.22) + 1.96 \times 0.0382 = -0.0052$$

$$L = (0.14 - 0.22) - 1.96 \times 0.0382 = -0.1546$$

Therefore, the 95 % confidence interval is,

$$(-0.1546 \leq p_1 - p_2 \leq -0.0052)$$

and the associated probability is

$$P(-0.1546 \leq p_1 - p_2 \leq -0.0052) = 0.95.$$

Finally, since the range from the lower limit to the upper limit is all negative numbers, the difference between the two proportions is significant at the 0.95 confidence level, and thereby, p_1 is significantly smaller than p_2.

Comparing k Means of Variable Type Data

Suppose a simulation model with terminating type data where the analyst is comparing various options in the model seeking the combination of options that yield the optimal efficiency. Could be a mixed model assembly line simulation model where the analyst is seeking the best way to sequence the units down the line to minimize the total idle time and congestion time on the line. One of the output measures is a variable type data. Each option is run with n repetitions and various outputs results are measured. The analyst wants to determine whether the difference in the output measure is significant.

The one-way analysis of variance is a method to test for significant differences in the output results. In the event, the options are found significantly different in an output measure, the next step is to determine which option(s) give significantly better results. Below shows how to use the one-way analysis of variance method.

One-Way Analysis of Variance

Assume k treatments and n repetitions of each are observed (in simulation runs). Note, in this situation, treatment is the same as option and is the common term in use for analysis of variance. When k treatments and n repetitions, the data available are:

$$x_{ij} \qquad i = 1, \ldots, k \text{ and } j = 1, \ldots, n.$$

The one-way analysis of variance method assumes each of the options, i, have mean μ_i and variance σ^2, all not known, and all are normally distributed. The null hypothesis is below:

$$Ho : \mu_1 = \ldots = \mu_k$$

The Type I error for this test is: $\alpha = P(\text{reject Ho} \mid \text{Ho is true})$.

In the event Ho is rejected, the analyst will seek the option(s) that yield significantly better results. One way to do this is by comparing the difference of two variables, as described earlier in this chapter. More of this is given subsequently.

A first step in using this method is to calculate the sample averages for each treatment, i, and for the total of all treatments as shown below.

$$\bar{x}_i = \sum_{j=1}^{n} x_{ij}/n \qquad i = 1 \text{ to } k$$

$$\bar{x} = \sum_{i=1}^{k} \bar{x}_l/k$$

A second step is to compute, the sum of squares of treatments, (SS_{TR}), and the sum of squares of error, (SS_E) as below.

$$SS_{TR} = n \sum_{i=1}^{k} (\bar{x}_i - \bar{x})^2$$

$$SS_E = \sum_{i=1}^{k} \sum_{j=1}^{n} (x_{ij} - \bar{x}_l)^2$$

The degrees of freedom for the treatments and for the error are the following:

$$df_{TR} = k - 1$$

$$df_E = nk - k$$

Next, the mean square for the treatments and the mean square of errors are obtained as follows:

$$MS_{TR} = SS_{TR}/df_{TR}$$

$$MS_E = SS_E/df_E$$

The residual errors for each observation are denoted as, $e_{ij} = (x_{ij} - \bar{x}_i)$, and the estimate of the variance becomes,

$$\hat{\sigma}^2 = MS_E$$

Further, the expected value of MS_E is

$$E(MS_E) = \sigma^2$$

The expected value of MS_{TR} depends on whether Ho is true or not, as below,

$$E(MS_{TR}) = \sigma^2 \text{ when Ho is true}$$

$$E(MS_{TR}) > \sigma^2 \text{ when Ho is not true}$$

To test if the null hypothesis is true, Fisher's test is applied and F_o is computed by,

$$F_o = MS_{TR}/MS_E$$

F_o has the pair of degrees of freedom (df_{TR}, df_E).

Next, the Fisher's F table is searched to find the value with confidence level α and degrees of freedom, (df_{TR}, df_E), denoted as $F\alpha_{(dfTR, dfE)}$. Finally, Ho is accepted or rejected depending on the outcome from below:

$$\text{If } F_o \leq F_{\alpha(dfTR, dfE)}, \text{ accept Ho}$$

$$\text{If } F_o > F_{\alpha(dfTR, dfE)}, \text{reject Ho}$$

Example 9.10 Suppose a simulation model is run with $k = 3$ options (treatments) and $n = 5$ repetitions for each option. One of the output measures is of the variable type and the smaller the value the better. The analyst is seeking whether any of the options yields significantly better results. The sample averages of the options and for the total are below:

Option i	Observation j					Average
	1	2	3	4	5	
1	10	9	11	8	12	$\bar{x}_1 = 10.0$
2	6	10	7	8	9	$\bar{x}_2 = 8.0$
3	9	5	6	9	6	$\bar{x}_3 = 7.0$
Total	$\bar{x} = 8.33$					

The associated sum of squares, degrees of freedom and mean squares are below.

Sum of squares	Degrees of freedom	Mean squares
$SS_{TR} = 23.35$	$df_{TR} = 2$	$MS_{TR} = 11.675$
$SS_E = 34.00$	$df_E = 12$	$MS_E = 2.833$

The measure of Fisher's F is

$$F_o = MS_{TR}/MS_E = 11.675/2.833 = 4.12$$

The F value in Fisher's tables with $\alpha = 0.05$ and degrees of freedom (df_{TR}, df_E) $= (2, 12)$ yields, $F_{0.05(2,12)} = 3.89$.

Since, $F_o > 3.89$, the null hypothesis is rejected, indicating there is a significant difference in the means from two or more of the options.

In this example, the smaller the mean, the better. The simulation results show where option 3 gives the best results and option 2 is the next best. The next step is to determine whether the sample mean values of options 2 and 3 are significantly different of not.

Example 9.11 Continuing with Example 9.10, the goal now is to determine whether the means of options 2 and 3 are significantly different. The estimate of the variance for the residual errors is $\hat{\sigma}^2 = MS_E = 2.833$; thereby, the standard error is $\hat{\sigma} = \sqrt{\hat{\sigma}^2} = 1.683$, and the associated degrees of freedom is $df_E = 12$. The $(1 - \alpha) = 95\%$ confidence limits (U and L) between the means of options 2 and 3 is computed using the student's t value with $\alpha/2 = 0.025$ and $df_E = 12$, whereby $t_{\alpha/2} = 2.179$. Hence,

$$U = (\bar{x}_3 - \bar{x}_2) + t_{\alpha/2}\hat{\sigma}/\sqrt{(2n)} = 0.160$$

$$L = (\bar{x}_3 - \bar{x}_2) - t_{\alpha/2}\hat{\sigma}/\sqrt{(2n)} = -2.160$$

The 95 % confidence interval becomes,

$$(-2.16 \leq \mu_3 - \mu_2 \leq 0.16)$$

and the corresponding probability is

$$P(-2.16 \leq \mu_3 - \mu_2 \leq 0.16) = 0.95$$

Because the values from L to U pass through zero, the mean of option 3 is not significantly smaller than the mean of option 2 with 95 % confidence.

Example 9.12 The 95 % confidence intervals of all three comparisons are below:

$$P(-2.16 \leq \mu_3 - \mu_2 \leq 0.16) = 0.95$$

$$P(-3.16 \leq \mu_2 - \mu_1 \leq -0.84) = 0.95$$

$$P(-4.16 \leq \mu_3 - \mu_1 \leq -1.84) = 0.95$$

The results show where option 1 is significantly higher from options 2 and 3 since the upper and lower limits are both negative and do not pass through zero. As stated, options 2 and 3 are not significantly different from each other, but option 1 is significantly higher. The analyst might consider taking more samples to gain further precision on comparing the difference between options 2 and 3.

Summary

This chapter describes the common statistical methods that are used to analyze the output data from computer models that are based on terminating and nonterminating systems. The statistical methods are essentially the same that are described in the common statistical textbooks. They include measuring the average, standard deviation, confidence interval from output data, some of the variable type and some of the proportion type. The methods described also pertain when the two or more variables are in review.

Chapter 10
Choosing the Probability Distribution from Data

Introduction

In building a simulation model, the analyst often includes several input variables of the control and random type. The control variables are those that are of the "what if" type. Often, the purpose of the simulation model is to determine how to set the control variables seeking optimal results. For example, in an inventory simulation model, the control variables may be the service level and the holding rate, both of which are controlled by the inventory manager. On each run of the model, the analyst sets the values of the control variables and observes the output measures to see how the system reacts.

Another type of variable is the input random variables, and these are of the continuous and discrete type. This type of variable is needed to match, as best as possible, the real life system for which the simulation model is seeking to emulate. For each such variable, the analyst is confronted with choosing the probability distribution to apply and the parameter value(s) to use. Often empirical or sample data is available to assist in choosing the distribution to apply and in estimating the associated parameter values. Sometimes two or more distributions may seem appropriate and the one to select is needed. The authenticity of the simulation model largely depends on how well the analyst can emulate the real system. Choosing the random variables and their parameter values is vital in this process.

This chapter gives guidance on the steps to find the probability distribution to use in the simulation model and how to estimate the parameter values that pertain. For each of the random variables in the simulation model with data available, the following steps are described: verify the data is independent, compute various statistical measures, choose the candidate probability distributions, estimate the parameter(s) for each probability distribution, and determine the adequacy of the fit.

N.T. Thomopoulos, *Essentials of Monte Carlo Simulation: Statistical Methods for Building Simulation Models*, DOI 10.1007/978-1-4614-6022-0_10,
© Springer Science+Business Media New York 2013

Collecting the Data

For each random variable in the simulation model, the analyst is obliged to seek actual data (empirical or sample) from the real system under study. For example, if the variable is the number of units purchased with each customer order for an item, the data collected might be the history of number of pieces from each order, for the item, in the past year, say. Since the numbers are integers, the variable is from a discrete probability distribution. The data is recorded and identified as $x_1, \ldots, x_n$, where n is the number of data entries collected. Subsequently, this sample or empirical data is the catalyst in selecting the probability distribution for the variable. The data is also needed to estimate the distribution's parameter value, and subsequently is applied to compare fitted values with the actual values.

Test for Independence

The data should be independently collected so that the subsequent statistical measures yield valid estimates. An example when not independent is the wait time for cars at a tollbooth line when the times are observed one car after the other. When the line is long, the successive wait times will remain high for the consecutive cars in a line. A way to avoid correlated data is to spread the samples out and not collect them one after the other. The common method to test for independent sequential data is to measure the autocorrelation with various lags.

Autocorrelation

Suppose the data, $x_1, \ldots, x_n$, is sequentially collected, and thereby may not be independent. A way to detect for independence is by measuring the autocorrelation of the data with various lags. The sample autocorrelation with a lag of k, denoted as r_k, is computed as below.

$$r_k = \sum_{i=k+1}^{n} \frac{(x_i - \bar{x})(x_{i-k} - \bar{x})}{\sum_{i=1}^{n}(x_i - \bar{x})^2} \quad k = 1, 2, 3, ..$$

where $\bar{x}$ is the average of all x's. When all sample autocorrelations are near zero, plus or minus, the data is assumed independent. In the event the data appears not independent, the sample should be retaken, perhaps sampling one item out of each five items, or one item per hour, so forth.

Example 10.1 Assume a series of observations are taken in sequence and the first three autocorrelations, say, are: $0.89, 0.54, 0.39$ for lags of $k = 1, 2, 3$, respectively. Since they are not near zero (plus or minus), the data does not appear as independent. On the other hand, if the first three autocorrelations were: $0.07, -0.13, 0.18$, the data does appear independent.

Some Useful Statistical Measures

A variety of statistical measures can be computed from the sample data, $x_1, \ldots, x_n$. Some that are useful in selecting the probability distribution are listed here:

$$x(1) = \text{minimum}$$

$$x(n) = \text{maximum}$$

$$\bar{x} = \text{average}$$

$$s = \text{standard deviation}$$

$$\text{cov} = s/\bar{x} = \text{coefficient of variation}$$

$$\tau = s^2/\bar{x} = \text{lexis ratio}$$

Example 10.2 Suppose 20 samples are the following: [5.3, 9.8, 5.1, 0.6, 3.9, 8.1, 4.0, 0.1, 4.6, 0.6, 2.9, 7.1, 2.7, 7.0, 2.5, 5.8, 3.0, 7.6, 3.5, 7.7]. The statistical measures from this data are listed below.

$$x(1) = 0.1$$

$$x(20) = 9.8$$

$$\bar{x} = 4.595$$

$$s = 2.713$$

$$\text{cov} = 0.590$$

$$\tau = 1.602$$

Location Parameter

Sometimes it is useful to estimate the location parameter for a distribution, labeled here as γ. This represents the minimum value of x, whereby, $x \geq \gamma$. Consider the sorted sample data denoted as $x(1) \leq x(2) \leq \ldots \leq x(n)$. A way to estimate γ for the Wiebull distribution is given by Zanakis (1979) and is below,

$$\hat{\gamma} = [x\,(1)\,x\,(n) - x(k)^2]/[x\,(1) + x(n) - 2x(k)]$$

where k is the smallest index with $x(k) > x(1)$.

Example 10.3 Suppose 50 observations are taken and are sorted as follows: [16.3, 21.3, 27.4, 35.7, 38.4, ..., 51.4], where 16.3 is the smallest and 51.4 the largest in the sample. Using the formula given above, the estimate of the minimum value of x becomes,

$$\hat{\gamma} = (16.3 \times 51.4 - 21.3^2)/(16.3 + 51.4 - 2 \times 21.3) = 15.3$$

Candidate Probability Distributions

The typical probability distribution candidates for a continuous random variable are the following: continuous uniform, normal, exponential, lognormal, gamma, beta and Wiebull. The more common discrete probability distributions are the discrete uniform, binomial, geometric, Pascal and Poisson.

Transforming Variables

In the pursuit of seeking the candidate distribution to use, it is sometimes helpful to convert a variable x to another variable, x', where x' ranges from zero to one, or where x' is zero or larger. More discussion is below.

Transform Data to (0,1)

A way to convert a variable to a range where x' lies between 0 and 1 is described here. Recall the summary statistics of the variable x as listed earlier. It is possible to estimate the summary statistics when the variable is transformed to lie between 0 and 1. For convenience in notation, let $a' = x(1)$ for the minimum, and $b' = x(n)$ for the maximum. When x is converted to x' by the relation $x' = (x - a')/(b' - a')$, the sample average and standard deviation become, $\bar{x}' = (\bar{x} - a')/(b' - a')$, and $s' = s/(b' - a')$, respectively. The corresponding coefficient of variation is $cov' = [s/(\bar{x} - a')]$. The cov of this measure may be useful when selecting the distribution to apply.

Transform Data to $(x \geq 0)$

A way to convert a variable to a range where x' lies approximately zero and larger is described here. Recall again the summary statistics of the variable x as listed earlier, and once more use the notation $a' = x(1)$ for the minimum. When x is converted to x' by the relation $x' = (x-a')$, the range of x' becomes zero or larger. The corresponding sample average and standard deviation become, $\bar{x}' = (\bar{x} - a')$, and $s' = s$, respectively. Finally, the coefficient of variation is $\text{cov}' = s/(\bar{x} - a')$.

Candidate Continuous Distributions

Below is a brief review on some of the properties of the continuous probability distributions. These are the following: continuous uniform, normal, exponential, lognormal, gamma, beta and Weibull. Of particular interest with each distribution is the coefficient of variation (cov) and its range of values that apply. When sample data is available, the sample cov can be measured and compared to each distribution's range to help narrow the choice for a candidate distribution.

Continuous Uniform

The random variable x from the continuous uniform distribution $(0,1)$ has a range of zero to one. The mean is $\mu = 0.5$ and standard deviation is $\sigma = 1/\sqrt{12} = 0.289$, and thereby, the coefficient of variation becomes $\text{cov} = \sigma/\mu = 0.577$.

Normal

When a variable x is normally distributed with mean μ and standard deviation σ, the notation is $x \sim N(\mu, \sigma^2)$. Note, the coefficient of variation for x is $\text{cov} = \sigma/\mu$. When all x values are zero or larger, the coefficient of variation is always 0.33 or smaller, i.e., $\text{cov} \leq 0.33$.

Exponential

Recall the exponential distribution where the variable x is zero or larger. The mean, μ, and standard deviation, σ, of this distribution have the same value and thereby, the coefficient of variation is $\text{cov} = \sigma/\mu = 1.00$.

Lognormal

When the variable x of a lognormal distribution is converted to the natural logarithm, ($x' = \ln(x)$), the notation for x is $x \sim LN(\mu', \sigma'^2)$, and for the transformation, it is $x' \sim N(\mu', \sigma'^2)$. Note, the parameters μ' and σ'^2, are the mean and variance, respectively, of x' the normal distributed variable and not the lognormal distributed variable. The coefficient of variation for the variable x' becomes $cov' = \sigma'/\mu'$.

Gamma

The variable x from the (standard) gamma distribution is always zero or larger and has parameters (k, θ). Recall, the mean and variance of x are $\mu = k/\theta$, and $\sigma^2 = k/\theta^2$, respectively, and therefore, the coefficient of variation is $cov = 1/\sqrt{k}$. When $k > 1$, cov is less than one. When $k \leq 1$, cov is one or larger. Note, the mode is $(k - 1)/\theta$ when $k \geq 1$, and is zero when $k < 1$.

Beta

The variable x from a beta distribution has many shapes that could skew right or left or be symmetric and look like the uniform, normal and may even have a bathtub-like shape. This distribution emulates most shapes, but is a bit difficult to apply. The parameters are (k_1, k_2), and the mean and variance are shown below:

$$\mu = \frac{k_1}{k_1 + k_2}$$

$$\sigma^2 = \frac{(k_1 k_2)}{(k_1 + k_2)^2 (k_1 + k_2 + 1)}$$

Weibull

The variable x from a Wiebull distribution has three parameters, (k_1, k_2, γ), where γ is the location parameter and can be estimated from the relation given earlier. The values of x are greater than γ and the shape is skewed to the right after the mode is reached. The mean and variance are below:

$$\mu = \frac{k_2}{k_1}\Gamma\left(\frac{1}{k_1}\right)$$

$$\sigma^2 = \frac{(k_2{}^2)}{k_1\left[2\Gamma\left(\frac{2}{k_1}\right) - 1/k_1\Gamma\left(\frac{1}{k_1}\right)^2\right]}$$

Some Candidate Discrete Distributions

An important statistic to determine the candidate discrete distribution is the lexis ratio, $\tau = \frac{\sigma^2}{\mu}$. The lexis ratio can be estimated from sample data by $\hat{\tau} = \frac{\hat{\sigma}^2}{\hat{\mu}}$, where $\hat{\mu} = \bar{x}$ = sample average, and $\hat{\sigma}^2 = s^2$ = sample variance. Below is a description on some of the properties concerning the lexis ratio for the more common discrete distributions.

Discrete Uniform

The variable x with the discrete uniform distribution has parameters, (a,b), where x are all the integers from a to b. The mean and variance of x are $\mu = (a + b)/2$ and $\sigma^2 = [(b - a + 1)^2 - 1]/12$, respectively. When $a = 0$, the lexis ratio is $\tau = \sigma^2/\mu = [(b + 1)^2 - 1]/6b$. Note, when $b \geq 4, \tau \geq 1$.

Binomial

The parameters for the binomial distribution are n (number of trials) and p (probability of a success per trial). The random variable is x (number of successes in n trials). The mean of x is $\mu = np$, and the variance is $\sigma^2 = np(1 - p)$. Hence, the lexis ratio, $\tau = \sigma^2/\mu = (1 - p) < 1$.

Geometric

Recall the geometric distribution where the parameter is p, the probability of a success on each trial. When the random variable is x (number of fails until the first success), $x = 0, 1, \ldots$, the mean is $\mu = (1 - p)/p$, the variance is $\sigma^2 = (1 - p)/p^2$, and the lexis ratio for x' becomes $\tau = \sigma^2/\mu = 1/p$ that is always larger than one.

But when x' = (x + 1) the variable is the number of trials until a success, x' = 1, 2, ..., the mean is $\mu = 1/p$, and the variance is $\sigma^2 = (1 - p)/p^2$. The lexis ratio, $\tau = \sigma^2/\mu = (1 - p)/p$, is inconclusive since the ratio ranges below and above one.

Pascal

The parameters for the Pascal distribution are p (probability of a success) and k (number of successes). The random variable is x (number of fails till k successes), where x = 0, 1, 2,..., the mean is $\mu = k(1 - p)/p$, and the variance is $\sigma^2 = k(1 - p)/p^2$. The lexis ratio is $\tau = \sigma^2/\mu = 1/p > 1$.

But when x' = (x + k) is the number of trials until k success's, x' = k, k + 1, ..., the mean is $\mu = k/p$, the variance remains as $\sigma^2 = k(1 - p)/p^2$, and the lexis ratio becomes $\tau = \sigma^2/\mu = (1 - p)/p$. In this situation, the lexis ratio ranges above and below one.

Poisson

The parameter for the Poisson distribution is θ (rate per unit of measure), where the unit of measure is typically a unit of time (minute, hour), and so forth. The random variable is x (number of events in a unit of measure). Since the mean of x is $\mu = \theta$, and the variance is $\sigma^2 = \theta$, the lexis ratio becomes $\tau = \sigma^2/\mu = 1$.

Estimating Parameters for Continuous Distributions

Below gives the popular ways to estimate the parameters for the common continuous distributions. These are by the maximum-likelihood estimators and/or the method-of-moment estimators.

Continuous Uniform

The parameters of the continuous uniform distribution are (a,b) where the variable x is equally likely to fall anywhere from a to b. When the data $x_1, ..., x_n$ is available, the maximum likelihood estimates of the parameters are as follows:

$$\hat{a} = \min(x_1, ..., x_n)$$

$$\hat{b} = \max(x_1, ..., x_n)$$

Another way to estimate the parameters for this distribution is by the method-of-moments. The same data is used to first compute the sample average, $\bar{x}$, and the sample standard deviation, s. Next, the estimates of the parameters are obtained in the following way:

$$\hat{a} = \bar{x} - \sqrt{12}s/2$$

$$\hat{b} = \bar{x} + \sqrt{12}s/2$$

Example 10.4 Consider a situation where the sample of n = 20 yield the following sorted data: [0.1, 0.6, 0.6, 2.5, 2.7, 2.9, 3.0, 3.5, 3.9, 4.0, 4.6, 5.1, 5.3, 5.8, 7.0, 7.1, 7.6, 7.7, 8.1, 9.8], and suppose the analyst suspects the data comes from a continuous uniform distribution and thereby needs estimates of the parameters, a and b. From the maximum likelihood estimator method, the estimates of the parameters are $\hat{a} = 0.1$ and $\hat{b} = 9.8$.

Another way to estimate the parameters is by the method-of-moments. To find the estimates this way, the average and standard deviation of the data entries are needed, and they are: $\bar{x} = 4.595$ and s = 2.713, respectively. Thereby the method-of-moment estimates become $\hat{a} = -0.10$ and $\hat{b} = 9.29$.

Normal Distribution

The normal distribution has two parameters, μ, the mean, and σ^2, the variance. The estimates are obtained from the sample mean, $\bar{x}$, and sample variance, s^2, as below,

$$\hat{\mu} = \bar{x}$$

$$\hat{\sigma}^2 = s^2$$

Example 10.5 Suppose the analyst has ten sample sorted data entries as [1.3, 6.4, 7.1, 8.7, 9.1, 10.2, 11.5, 14.3, 16.1, 18.0]. The sample average is $\bar{x} = 10.27$ and the standard deviation is s = 4.95. Hence, x is estimated as: $N(10.27, 4.95^2)$.

Exponential

The exponential distribution has one parameter, θ, where the mean and standard deviation of x are equal whereby $\mu = \sigma = \frac{1}{\theta}$. The maximum-likelihood-estimator of the parameter is based on the sample mean, $\bar{x}$, as shown below,

$$\hat{\theta} = 1/\bar{x}$$

Example 10.6 Suppose the analyst has the following data with $n = 10$ observations: 3.0, 5.7, 10.8, 0.3, 1.5, 2.5, 4.5, 7.3, 1.3, 2.1, and assumes the data comes from an exponential distribution. The sample average is $\bar{x} = 3.90$, and thereby the estimate of the exponential parameter is $\hat{\theta} = 1/\bar{x} = 1/3.90 = 0.256$. Upon further computations, the standard deviation of the ten observations is measured as $s = 3.24$, not too far away from the average of 3.90.

Lognormal

Consider the variable x of the lognormal distribution, and another related variable, y, that is the natural logarithm of x, i.e. $y = \ln(x)$. The parameters for x are the mean and variance of y and are denoted as, μ_y, and σ_y^2, respectively. To estimate the parameters, the n corresponding values of y $(y_1, \ldots, y_n)$ are needed to give the sample average, $\bar{y}$, and the sample variance, s_y^2. The estimates of the parameters for the lognormal distribution are the following:

$$\hat{\mu}_y = \bar{y}$$

$$\hat{\sigma}_y^2 = s_y^2$$

Example10.7 Assume the analyst has collected ten sample entries as X = [0.3, 1.3, 1.5, 2.1, 2.5, 3.0, 4.5, 5.7, 7.3, 10.8]. Upon taking the natural logarithm of each, $y = \ln(x)$, the sample now has ten variables on y. The corresponding values of y are Y = [−1.204, 0.262, 0.405, 0.742, 0.916, 1.099, 1.504, 1.741, 1.988, 2.379]. The mean and variance of the n = 10 observations of y are $\bar{y} = 0.983$ and $s_y^2 = 1.057$.

Gamma

The variable x from the gamma distribution has two parameters (k, θ). The mean of x is $\mu = k/\theta$, and the variance is $\sigma^2 = k/\theta^2$. One way to estimate the parameters is by the method-of-moments using the sample average, $\bar{x}$, and the sample variance, s^2 that are computed from data, $x_1, \ldots, x_n$. The estimate of the gamma parameters are derived from,

$$\hat{\theta} = \bar{x}/s^2$$

$$\hat{k} = \bar{x}\hat{\theta}$$

Example 10.8 Assume a sample of n entries, $x_1, \ldots, x_n$, are collected, from which the average and variance are measured as $\bar{x} = 10.8$ and $s^2 = 4.3$, respectively. The analyst wants to estimate the gamma parameters for this data. Using the method of moments, the estimates are: $\hat{\theta} = 2.51$ and $\hat{k} = 27.12$.

Beta

The variable x from the beta distribution (0–1) has two parameters (k_1, k_2). The mean of x is $\mu = \frac{k_1}{k_1 + k_2}$, and when the two parameters are greater than zero, $(k_1 > 0,$ $k_2 > 0)$, the mode is $\tilde{\mu} = \frac{(k_1 - 1)}{(k_1 + k_2 - 2)}$. In the typical situation, the distribution skews to the right. This occurs when $k_2 > k_1 > 1$. For this situation, a way to estimate the parameters is with use of the sample average, $\bar{x}$, and the sample mode, $\tilde{x}$. From the two equations and two unknowns, and some algebra, the estimates of the parameters are computed as below:

$$\hat{k}_1 = \bar{x}[2\tilde{x} - 1]/[\tilde{x} - \bar{x}]$$

$$\hat{k}_2 = [1 - \bar{x}]\hat{k}_1/\bar{x}$$

Example 10.9 Assume sample data of x that lies between 0 and 1, and yield the average and mode as $\bar{x} = 0.4$ and $\tilde{x} = 0.2$, respectively. The analyst wants to estimate the parameters for a beta distribution in the range (0–1). The estimates are below:

$$\hat{k}_1 = 0.4[2 \times 0.2 - 1]/[0.2 - 0.4] = 1.2$$

$$\hat{k}_2 = [1 - 0.4]1.2/[0.4] = 1.8$$

Estimating Parameters for Discrete Distributions

Below gives the popular ways to estimate the parameters for the common discrete distributions. These are by the maximum-likelihood estimators and/or the method-of-moment estimators.

Discrete Uniform

The variable x from the discrete uniform distribution has two parameters (a,b) where the variable x is equally likely to fall as an integer from a to b. The sample data $(x_1, \ldots, x_n)$ is used to find the minimum, $x(1)$, and maximum, $x(n)$. The maximum likelihood estimator of the parameters, a and b, are obtained as below:

$$\hat{a} = x(1)$$

$$\hat{b} = x(n)$$

Another way to estimate the parameters is by the method-of-moments. The mean of x is $\mu = (a + b)/2$ and the variance is $\sigma^2 = [(b-a-1)^2-1]/12$. Using the sample mean $\bar{x}$, and sample variance, s^2, and a bit of algebra, the following parameter estimates are found:

$$\hat{a} = \text{floor integer of } (\bar{x} + 0.5 - 0.5\sqrt{12s^2 + 1})$$

$$\hat{b} = \text{ceiling integer of } (\bar{x} - 0.5 + 0.5\sqrt{12s^2 + 1})$$

Example 10.10 Suppose an analyst collects ten discrete sample data [7, 5, 4, 8, 5, 4, 12, 9, 2, 8] and wants to estimate the min and max coefficients from a discrete uniform distribution. Using the maximum likelihood estimator, the minimum and maximum estimates are:

$$\hat{a} = 2$$

$$\hat{b} = 12$$

The method-of-moment estimate of the parameters requires finding the sample average and sample variance. These are: $\bar{x} = 6.4$ and $s^2 = 8.711$. So now, the estimate of the parameters become,

$$\hat{a} = \text{floor } (6.4 + 0.5 - 0.5\sqrt{12 \times 8.711 + 1}) = \text{floor } (1.764) = 1$$

$$\hat{b} = \text{ceiling } (6.4 - 0.5 + 0.5\sqrt{12 \times 8.711 + 1}) = \text{ceiling}(11.036) = 12$$

Binomial

The variable x from the binomial distribution has parameters (n, p), where typically n is known and p is not. The expected value of x is $E(x) = np$, and thereby, when a sample of n trials yields x successes, the maximum likelihood estimate of p is,

$$\hat{p} = x/n.$$

In the event the n trial experiment is run m times, and the results are $(x_1, \ldots, x_m)$, with an average of $\bar{x}$, the estimate of p becomes,

$$\hat{p} = \bar{x}/n.$$

Example 10.11 Suppose m = 5 experiments of binomial data with n = 8 trials are observed with the results: [1, 3, 2, 2, 0]. Since the average is $\bar{x} = 1.6$, the estimate of p is $\hat{p} = 1.6/8 = 0.2$.

Geometric

Consider the geometric distribution where the variable x (0, 1, 2, ...) is the number of fails before the first success and p is the probability of a success per trial. The expected value of x is $E(x) = (1-p)/p$. When m samples of x are taken, with results $(x_1, \ldots, x_m)$, and a sample average $\bar{x}$, the maximum-likelihood-estimator of p becomes,

$$\hat{p} = 1/(\bar{x} + 1).$$

Example 10.12 Suppose m = 8 samples from geometric data are observed and yield the following values of x: [3,6, 2, 5, 4, 4, 1, 5] where x is the number of failures till the first success. The analyst wants to estimate the probability of a success, p, and since the average of x is $\bar{x} = 3.75$, the estimate becomes $\hat{p} = 1/(3.75 + 1) = 0.211$.

Pascal

Recall the Pascal distribution where the variable x is the number of failures till k success's. The parameters are (k, p), where k is known, and assume p is not known. When m samples of x are taken, with results $(x_1, \ldots, x_m)$, and a sample average $\bar{x}$ is computed, the maximum-likelihood-estimator of p becomes the following:

$$\hat{p} = k/(\bar{x} + k).$$

Example 10.13 Suppose m = 5 samples from the Pascal distribution with parameter k = 4 are observed and yield the following data entries of x: [6, 4, 7, 5, 6] where x is the number of failures till k successes. The analyst wants to estimate the probability of a success, and since the average of x is $\bar{x} = 5.60$, the estimate is $\hat{p} = 4/(5.60 + 4) = 0.417$.

Poisson

The variable x (0, 1, 2, ...) from the Poisson distribution has parameter θ. The expected value of x is $E(x) = \theta$. When m samples of x $(x_1, \ldots, x_m)$ are collected, the sample average of x is readily computed as $\bar{x}$. Using the sample mean, the maximum likelihood estimator of θ becomes,

$$\hat{\theta} = \bar{x}.$$

Example 10.14 Suppose m = 10 samples from Poisson data are observed and yield the following values of x: [0, 0, 1, 2, 2, 0, 1, 2, 0, 1]. The analyst wants to estimate the Poisson parameter, θ, and since the average of x is $\bar{x} = 0.90$, the estimate is $\hat{\theta} = 0.90$.

Q-Q Plot

The Q-Q plot is a graphical way to compare the quantiles of sample (empirical) data to the quantiles from a specified probability distribution as a way of observing the goodness-of-fit. This plot applies to continuous probability distributions. See Wilk and Ganandel (1968) for a fuller description on the Q-Q (quantile to quantile) plot. To carryout, the empirical or sample data [x_1,, ..., x_n] are first arranged in sorted order [x(1),. ..., x(n)] where x(1) is the smallest value and x(n) the largest. The quantiles for the sample data are merely [x(1), ..., x(n)]. The empirical cumulative distribution function (cdf) of the sample quantiles is computed and denoted as

$$F[x(i)] = w_i = (i - 0.5)/n \qquad i = 1 \text{ to } n$$

For example if n = 10, and i = 1, F[x(1)] = w_1 = 0.05. At i = 2, F[x(2)] = w_2 = 0.15; at i = 10, F[x(10)] = w_{10} = 0.95, and so forth. The set of ten probabilities are denoted as Ps = [w_1, ..., w_{10}]. Note, for each x(i), there is an associated w_i.

Consider a probability distribution, f(x) where the cumulative probability distribution is F(x). The corresponding quantiles for this distribution are obtained by the inverse function,

$$x_i' = F^{-1}(w_i) \qquad i = 1 \text{ to } n$$

For each quantile from the sample, a corresponding quantile is computed for the probability distribution. For convenience, the pair of quantiles are labeled as Xs = [x(1), ..., x(n)] for the sample data, and Xf = [x_1', ..., x_n'] for the fit from the probability model.

The n pair of quantiles are now placed on a scatter plot with the sample quantiles, Xs, on the x-axis and the probability model quantiles, Xf, on the y-axis. In the event the probability model is a good fit to the sample data, the scatter plot will look like a straight line going through a $45°$ angle from the lower left-hand side to the upper right-hand side, and the scale of the x and y axis will be similar. In the literature, it is noted where some references place Xs on the y-axis and Xf on the x-axis.

Example 10.15 Suppose n =5 sample (or empirical)data of a variable are observed as: [8.3, 2.5, 1.3, 9.4, 5.0]. The data are sorted and the sample quantiles are: Xs = [1.3, 2.5, 5.0, 8.3, 9.4]. The set of empirical probabilities are obtained from the n samples and are listed in vector form as: Ps = [0.1, 0.3, 0.5, 0.7, 0.9].

Assume the sample data are to be compared to a continuous uniform distribution where $f(x) = 0.1$ for $0 \leq x \leq 10$. Since the cumulative distribution of x becomes $F(x) = 0.1x = w$, the quantile for each w is obtained by $x = F^{-1}(w) = w/0.1$. For each probability on the sample set, Ps, an associated fit from the model quantile is computed as

$$x_i = w_i/0.1 \quad i = 1 \text{ to } 5.$$

Thereby, the five probability fit quantiles are $Xf = [1.0, 3.0, 5.0, 7.0, 9.0]$. The Q-Q plot for the pair of quantiles (Xs, Xf) is shown in Fig. 10.1. Since the scatter appears much like a straight line with a 45° fit, the conclusion is that the sample data is a reasonably close fit to the continuous uniform distribution that is under consideration.

Example 10.16 Consider once more the sample data from Example 10.15 where $n = 5$ and the sorted data yield the quantile set: $Xs = [1.3, 2.5, 5.0, 8.3, 9.4]$, and associated empirical probabilities $Ps = [0.1, 0.3, 0.5, 0.7, 0.9]$.

Now suppose the sample data are to be compared to a continuous distribution, f $(x) = x/50$ for $0 \leq x \leq 10$. Note, the cumulative distribution of x becomes $F(x) = x^2/100 = w$. So now, for probability w, the quantile is computed by $x = F^{-1}(w) = \sqrt{100F(x)} = 10\sqrt{w}$. Hence, for each w in the sample set, Ps, an associated fitted quantile is obtained by,

$$x_i = 10\sqrt{w_i} \quad i = 1 \text{ to } 5.$$

Thereby, the five quantiles for the probability fit become $Xf = [3.2, 5.5, 7.1, 8.4, 9.5]$. The Q-Q plot for the pair of quantiles (Xs, Xf) is shown in Fig. 10.2. Since the scatter plot is not close to the 45° line, the conclusion is that the sample data is not a good fit to the probability distribution under consideration.

Example 10.17 (Continuous Uniform Q-Q Plot). Consider the sample data from Example 10.4, and suppose the analyst wants to run a Q-Q Plot assuming the probability distribution is a continuous uniform distribution. Recall from the earlier example, the MLE estimates of the parameters are $\hat{a} = 0.1$ and $\hat{b} = 9.8$. Hence, the probability function is estimated as $f(x) = 1/9.7$ for $0.1 \leq x \leq 9.8$, and the cumulative distribution is $F(x) = (x-0.1)/9.7$. So, when the cumulative probability is $w = F(x)$, the associated quantile becomes $w = F^{-1}(w) = 0.1 + w(9.7)$. At $i = 1$, the minimum rank, $w_i = (1-0.5)/20 = 0.025$ and $x_1' = 0.1 + 0.025$ $(9.7) = 0.345$. At $i = 20$, the maximum rank, $w_{20} = (20-0.5)/20 = 0.975$ and $x_{20}' = 0.1 + 0.975(9.7) = 9.56$, and so forth. The full set of quantiles for the probability fit is denoted as Xf, and for simplicity, are listed here with one decimal accuracy, $Xf = [0.3, 0.8, 1.3, 1.8, 2.3, 2.8, 3.3, 3.7, 4.2, 4.7, 5.2, 5.7, 6.2, 6.7, 7.1, 7.6, 8.1, 8.6, 9.1, 9.6]$.

Using the pair (Xs, Xf), the Q-Q Plot is in Fig. 10.3. The vector Xs contains the sample data from Example 10.4. Note, the scatter plot closely fits a 45° angle from

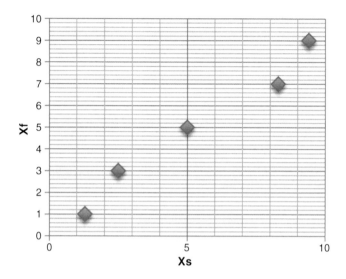

Fig. 10.1 Q-Q plot when $f(x) = 0.1$ for $0 \leq x \leq 10$

the lower left corner to the upper right corner, and thereby the sample data appears as a good fit to the continuous distribution with parameters $\hat{a} = 0.1$ and $\hat{b} = 9.8$.

Example 10.18 (Normal Q-Q Plot). Recall Example 10.5 where an analyst has collected a sample of $n = 10$ observations: [7.1, 9.1, 11.5, 16.1, 18.0, 14.3, 10.2, 8.7, 6.4, 1.3],and where the sample average and standard deviation of the data are: $\bar{x} = 10.27$ and $s = 4.95$, respectively. The sorted values gives the sample quantile set Xs = [1.3, 6.4, 7.1, 8.7, 9.1, 10.2, 11.5, 14.3, 16.1, 18.0]. With $n = 10$, the ith sorted cumulative probability for the sample quantiles are $w_i = (i - 0.5)/10$ for ($i = 1$ to 10). The set Ps of cumulative probabilities is: Ps = [0.05, 0.15, 0.25, 0.35, 0.45, 0.55, 0.65, 0.75, 0.85, 0.95]. Note the cumulative probabilities for $i = 1$ is $w_1 = 0.05$, for $i = 2$, it is $w_2 = 0.15$, and so forth.

The analyst wishes to explore how the data fits with the normal distribution. To do this, the z variable from the standard normal distribution is needed with each w entry in the sample probability set Ps. This gives another set here denoted as Z = [−1.645,−1.036,−0.674,−0.385,−0.125, 0.125, 0.385, 0.674, 1.036, 1.645]. See Table A.1 in the Appendix. Note at $i = 1$, $w_1 = 0.05$ and $z_1 = -1.645$, whereby $P(z < -1.645) = 0.05$. In the same way, all the z values are obtained from the standard normal distribution. Now using the average, $\bar{x}$, standard deviation, s, and z values, it is possible to compute the $n = 10$ fitted quantiles for the normal distribution by the following formula,

$$x_i' = \bar{x} + z_i s \qquad i = 1 \text{ to } 10$$

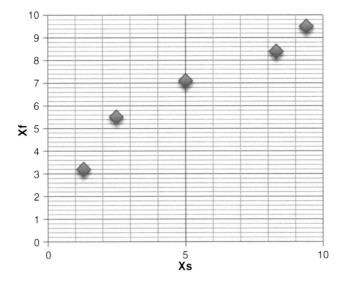

Fig. 10.2 Q-Q plot when f(x) = x/50 for $0 \leq x \leq 10$

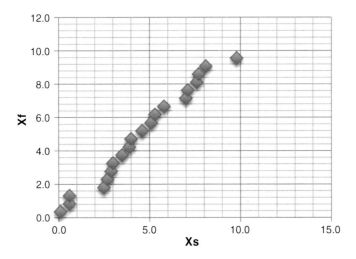

Fig. 10.3 Q-Q plot for the continuous uniform example

Applying the above formula yields the quantiles for the probability model, Xf = [2.12, 5.14, 6.93, 8.36, 9.65, 10.89, 12.18, 13.61, 15.40, 18.42]. The Q-Q Plot comparing the quantiles from the sample, Xs, with the quantiles from the probability fit, Xf, is shown in Fig. 10.4. Since the plot closely follows the 45° line from the lower left-hand side to the upper right-hand side, the sample data seems like a good fit with the normal distribution.

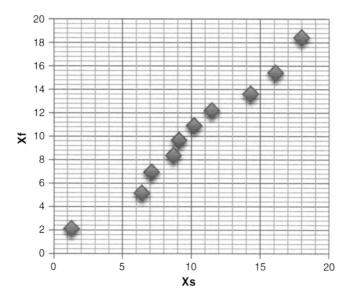

Fig. 10.4 Q-Q plot for the normal example

Example 10.19 (Exponential Q-Q Plot). Recall the n $=$ 10 observations from Example 10.6 where the sample average and standard deviation were $\bar{x} = 3.90$ and $s = 3.24$, respectively. The sorted data yields the sample quantiles, Xs $=$ [0.3, 1.3, 1.5, 2.1, 2.5, 3.0, 4.5, 5.7, 7.3, 10.8]. Since n $=$ 10, the ith sorted cumulative probability for the sample quantiles are $w_i = (i - 0.5)/10$ for (i $=$ 1 to 10), and the set Ps of cumulative probabilities is: Ps $=$ [0.05, 0.15, 0.25, 0.35, 0.45, 0.55, 0.65, 0.75, 0.85, 0.95].

Now assume the analyst wants to compare the data to an exponential distribution with parameter $\theta = 1/\bar{x} = 1/3.9 = 0.256$. The exponential density is $f(x) = \theta \exp(-\theta x)$, and the cumulative distribution is, $F(x) = 1 - \exp(-\theta x)$. For a given cumulative probability $w = F(x)$, the associated value of x is obtained by the relation below.

$$x = -1/\theta \ln(1 - w)$$

where ln is the natural logarithm.

So for the n $=$ 10 values of w listed above, the corresponding values of x are computed and are the ten fitted quantiles for the exponential distribution. They are labeled as Xf, whereXf $=$ [0.20, 0.63, 1.12, 1.68, 2.33, 3.11, 4.09, 5.41, 7.40, 11.68]. See Fig. 10.5 showing the Q-Q Plot that relates the sample quantiles, Xs, with the exponential quantiles, Xf, is below. Because the plot is a good fit through a 45° line from the lower left-hand corner to the upper right-hand corner, the exponential distribution appears as a good fit to the ten sample observations.

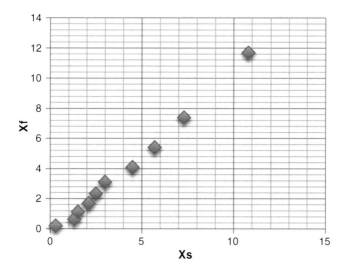

Fig. 10.5 Q-Q plot for exponential example

Example 10.20 (Lognormal Q-Q Plot) Suppose the analyst wants to run a Q-Q Plot comparing the lognormal distribution on the same data of Example 10.7. Recall, the ten observations are Xs = [0.3, 1.3, 1.5, 2.1, 2.5, 3.0, 4.5, 5.7, 7.3, 10.8], and the cumulative probabilities are: Ps = [0.05, 0.15, 0.25, 0.35, 0.45, 0.55, 0.65, 0.75, 0.85, 0.95]. This gives the set Z = [−1.645, −1.036, −0.674, −0.385, −0.125, 0.125, 0.385, 0.674, 1.036, 1.645]. Recall at i = 1, w_1 = 0.05 and z_1 = −1.645, whereby P(z < −1.645) = 0.05, and so forth. All the z values are obtained from the standard normal distribution. To test for the lognormal, the natural logarithm of each sample is taken as y_i = ln(x_i) for i = 1–10, The ten quantiles are the transformed data and are denoted as Ys = [−1.204, 0.262, 0.405, 0.742, 0.916, 1.099, 1.504, 1.741, 1.988, 2.379]. The average and standard deviation on the ten values of y are, $\bar{y}$ = 0.9833 and s = 1.0283, respectively. For each z_i, the corresponding(fitted) entry is obtained from the relation below,

$$y_i' = \bar{y} + z_i s \qquad\qquad \text{for i = 1 to 10.}$$

The ten fitted values of the normal distribution are now compared to their counterpart y_i (i = 1 to 10), and are listed as Yf = [−0.706, −0.080, 0.292, 0.589, 0.857, 1.113, 1.381, 1.678, 2.051, 2.677]. The ten-paired data of Ys and Yf form the Q-Q Plot in Fig. 10.6. Since the plotted data lie below the 45° line from the lower left-hand corner to the upper right-hand corner, the lognormal distribution does not appear as a good fit for the data.

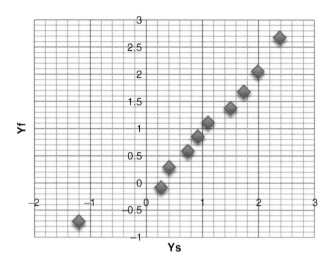

Fig. 10.6 Q-Q plot for the lognormal example

P-P Plot

The P-P plot is a graphical way to compare the cumulative distribution function (cdf) of sample (or empirical) data to the cdf of a specified probability distribution as a way of detecting the goodness-of-fit. The plot applies for both continuous and discrete probability distributions. See Wilk and Ganandel (1968) for a fuller description on the P-P (probability to probability) plot. To carryout, the sample data $(x_1,, \ldots, x_n)$ are first arranged in sorted order $[x(1),. \ldots, x(n)]$ where $x(1)$ is the smallest value and $x(n)$ the largest. The cdf for the sample data, $F[x(i)]$ $i = 1$ to n, are denoted here as $(w_1, \ldots, w_n)$, where

$$w_i = F[x(i)] = (i - 0.5)/n \qquad i = 1 \text{ to n}$$

For example if n $= 10$, and i $= 1, F[(1)] = w_1 = 0.05$. At i $= 2, F[x(2)] = 0.15$; at i $= 10, F[x(10)] = w_{10} = 0.95$, and so forth. The set of ten probabilities are denoted as Fs $= [w_1, \ldots, w_{10}]$. Note, there is one w_i for each $x(i)$.

Now consider a probability distribution, $f(x)$ where the cumulative probability distribution is $F(x)$. The corresponding cdf's for this distribution are obtained by

$$w_i' = F(x_{(i)}) \qquad i = 1 \text{ to n}$$

For each cdf from the sample, a corresponding cdf is computed for the probability distribution. For convenience, the pair of cdf's are labeled as Fs $= [w_1, \ldots, w_{10}]$ for the sample data, and Ff $= [w_1', \ldots, w_{10}']$ for the probability model.

The n pair of cdf's are now placed on a scatter plot with the sample cdf's, Fs, on the x-axis and the probability model cdf's, Ff, on the y-axis. In the event the

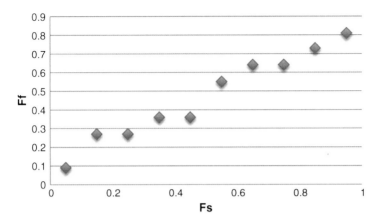

Fig. 10.7 P-P plot for discrete uniform example

probability model is a good fit to the sample data, the scatter plot will look like a straight line going through a 45° angle from the lower left hand side to the upper right hand side, and the scale of the x-axis and y-axis are similar. Note, in some references, Fs is placed on the y-axis and Ff on the x-axis.

Example 10.21 (Discrete Uniform) Recall Example 10.10 where n $=$ 10 samples were taken from data assumed as discrete uniform, and where the maximum likelihood estimates of the parameters are $\hat{a} = 2$ and $\hat{b} = 12$. The sorted data becomes [2, 4, 4, 5, 5, 7, 8, 8, 9, 10]. Since n $=$ 10, the sample cdf for this data are listed here as Fs $=$ [0.05, 0.15, 0.25, 0.35, 0.45, 0.55, 0.65, 0.75, 0.85, 0.95]. Recall, for a discrete uniform distribution, with parameters (a,b), the cdf is computed by F[x] $=$ (x $-$ a+1)/(b $-$ a+1). Hence, the cdf's for the fitted probability model become: Ff $=$ [0.09, 0.27, 0.27, 0.36, 0.36, 0.55, 0.64, 0.64, 0.73, 0.81]. The P-P Plot is shown in Fig. 10.7. Since the plotted points are similar to a 45° line, the sample data appears reasonably close to a discrete uniform distribution.

Example 10.22 (Binomial) Consider Example 10.11 where m $=$ 5 samples on binomial data for the number of successes in n $=$ 8 trials yields an estimate of $\hat{p} = 0.20$. The sorted data is Xs $=$ [0, 1, 2, 2, 3] and the cdf of the sample data becomes: Fs $=$ [0.1, 0.3, 0.5, 0.7, 0.9]. From the binomial distribution, the probability of x successes in n $=$ 8 trials with p $=$ 0.2 are computed as: p(0) $=$ 0.168, p(1) $=$ 0.336, p(2) $=$ 0.293, p(3) $=$ 0.146, ... Hence the associated cdf's for the fitted probability model are: F(0) $=$ 0.168, F(1) $=$ 0.504, F(2) $=$ 0.797, F(3) $=$ 0.943, ..., and thereby Ff $=$ [0.168, 0.504, 0.797, 0.797, 0.943]. Figure 10.8 is the P-P Plot for this data. The plot somewhat follows a 45° line and as such, the binomial probability distribution appears as a fair fit to the data.

Example 10.23 (Geometric) Recall Example 10.12 where m $=$ 8 samples are taken from a geometric distribution where x is the number of failures till a success. The estimate of the success probability for the example is $\hat{p} = 0.211$. The sorted

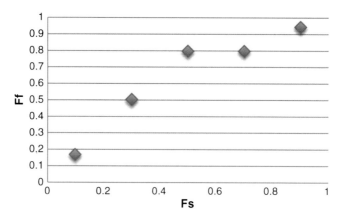

Fig. 10.8 P-P plot for binomial example

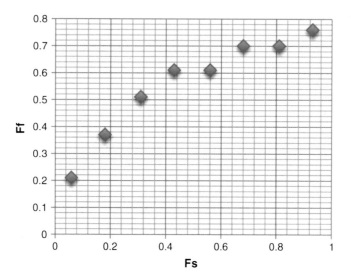

Fig. 10.9 P-P plot for the geometric example

data are Xs = [1, 2, 3, 4, 4, 5, 5, 6], and the corresponding cdf's are Fs = [0.0625, 0.1875, 0.3125, 0.4375, 0.5625, 0.6875, 0.8125, 0.9375]. Using p = 0.211, and the cumulative distribution function,

$$F(x) = 1 - (1 - p)^{x+1} \qquad x = 0, 1, \ldots$$

Ff = [0.377, 0.508, 0.612, 0.694, 0.694, 0.758, 0.758, 0.809]. Figure 10.9 is the P-P Plot for this example. Since the plot does not follow a 45° line, the geometric distribution does not appear as a good fit to the data.

Fig. 10.10 P-P plot for poisson example

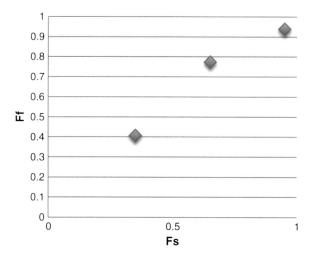

Example 10.24 (Poisson) Consider Example 10.14 with m = 10 samples on Poisson data where the parameter estimate is $\hat{\theta} = 0.90$ and the sorted sample data is Xs = [0, 0, 0, 0, 1, 1, 1, 2, 2, 2]. The cdf of the sample data is Fs = [0.05, 0.15, 0.25, 0.35, 0.45, 0.55, 0.65, 0.75, 0.85, 0.95]. Applying the Poisson probability with $\theta = 0.90$, the probabilities of x are: p(0) = 0.407, p(1) = 0.366, p(2) = 0.165, ..., which yields the cdf for the fit as Ff = [0.407, 0.407, 0.407, 0.407, 0.773, 0.773, 0.773, 0.938, 0.938, 0.938, ...].

Adjustment for Ties

In this example, there are (many) ties in the sample data, and special consideration is applied in the P-P Plot. With ties in the sample (or empirical) data, the P-P plot only considers one entry for each unique value of the data. In this way, Xs = [0, 1, 2], Fs = [0.35, 0.65, 0.95], and Ff = [0.407, 0.773, 0.938]. Figure 10.10 is the P-P Plot for the cdf's of the sample and the fit. Since, Fs is similar to Ff, the fit seems appropriate.

Summary

Computer simulation models often include one or more variables that play important roles in the model. Some of the random variables are of the continuous type and others are discrete. The analyst is confronted with choosing the proper probability distribution for each variable, and also with estimating the associated parameter(s) value. The chapter describes some of the common ways to select the distribution and to estimate the associated parameter values when some empirical or sample data is available from the real system.

Chapter 11
Choosing the Probability Distribution When No Data

Introduction

Sometimes the analyst has no data to measure the parameters on one or more of the input variables in a simulation model. When this occurs, the analyst is limited to a few distributions where the parameter(s) may be estimated without empirical or sample data. Instead of data, experts are consulted who give their judgment on various parameters of the distributions. This chapter explores some of the more common distributions where such expert opinions are useful. The distributions described here are continuous and are the following: continuous uniform, triangular, beta, lognormal and Weibull. The data provided by the experts is the following type: minimum value, maximum value, most likely value, average value, and a p-quantile value.

Continuous Uniform

Recall the continuous uniform distribution, CU (a,b), with parameters, a and b, where the variable x is equally likely to fall anywhere from a to b. The probability distribution is $f(x) = 1/(b-a)$, and the cumulative distribution is $F(x) = (x - a)/(b - a)$ for $a \leq x \leq b$. Sometimes the analyst wants to use this distribution but does not have data to estimate the parameters, (a, b). Suppose expert(s) can help by providing their opinions on the range of the statistics as below.

First assume the expert(s) can give the following two estimates on the distribution:

$\hat{a}$ = an estimate of the minimum value of x, and x_α = an estimate of the of α-quantile of x where $P[x \leq x_\alpha] = \alpha$. An estimate of the parameter b is needed to use the distribution in the simulation model. Using the estimates provided, the cumulative distribution becomes,

$$\alpha = F(x) = (x_\alpha - \hat{a})/(b - \hat{a}),$$

N.T. Thomopoulos, *Essentials of Monte Carlo Simulation: Statistical Methods for Building Simulation Models*, DOI 10.1007/978-1-4614-6022-0_11, © Springer Science+Business Media New York 2013

and thereby the estimate of b is,

$$\hat{b} = \hat{a} + (x_\alpha - \hat{a})/\alpha$$

Example 11.1 Suppose the analyst wants to use the continuous uniform distribution in the simulation model and has estimates of the minimum value of x as $\hat{a} = 10$ and the 0.9-quantitleas $x_{0.9} = 15$. With this information, the estimate of b becomes,

$$\hat{b} = 10 + \frac{15 - 10}{0.90} = 15.56.$$

So now, the distribution to use is,

$$f(x) = 1/5.56 \quad \text{for } 10.00 \le x \le 15.56$$

Now assume the expert(s) can give the following two estimates on the distribution:

$\hat{b}$ = an estimate of the maximum value of x, and x_α = an estimate of the of α-quantile of x where $P[x \le x_\alpha] = \alpha$. An estimate of the parameter a is needed to use the distribution in the simulation model. Using the estimates provided, the cumulative distribution becomes,

$$\alpha = F(x) = (x_\alpha - a)/(\hat{b} - a),$$

and thereby, using $\hat{b}$ and x_α, and some algebra, the estimate of the minimum parameter a is,

$$\hat{a} = (x_\alpha - \alpha\hat{b})/(1 - \alpha)$$

Example 11.2 Assume a situation where the simulation analyst wants to use the continuous uniform distribution and has estimates of $\hat{b} = 16$ and the 0.1-quantile $x_{0.1} = 11$. The estimate of a becomes,

$$\hat{a} = (11 - 0.1 \times 16)/(1 - 0.1) = 10.44$$

So now, the distribution to use is,

$$f(x) = 1/5.56 \quad \text{for } 10.44 \le x \le 16.00$$

The following sections show how to apply the triangular, beta, Weibull and lognormal distributions in the simulation model when no data is available. See Law, pages 370–375 (2007) for further discussion.

Triangular

Recall, the triangular distribution applies for a continuous variable, x with three parameters, $(a, b, \tilde{x})$, where the range of x is from a to b, and the mode is denoted as $\tilde{x}$. When the analyst wants to use this distribution in a simulation model and has no empirical or sample data to estimate the three parameters, he/she may turn to one or more experts to gain the estimates of the following type:

$\hat{a}$ = an estimate of the minimum value of x

$\hat{b}$ = an estimate of the maximum value of x

$\hat{\tilde{x}}$ = an estimate of the most likely value of x

So now, the triangular distribution can be used with parameters, $\hat{a}, \hat{b}, \hat{\tilde{x}}$.

The associated standard triangular distribution, $T(0, 1, \tilde{x}')$, with variable x' falls in the range from 0 to 1. The most likely value of x' is the mode denoted as $\tilde{x}'$. The mode of the standard triangular variable is computed from the corresponding triangular variable by $\tilde{x}' = (\tilde{x} - \hat{a})(\hat{b} - \hat{a})$.

Example 11.3 Suppose the analyst wants to use the continuous triangular distribution in the simulation model and from experts opinions has estimates of $\hat{a} = 10$, $\hat{b} = 60$ and $\hat{\tilde{x}} = 20$. To apply the standard triangular, with variable x', the estimate of the mode becomes:

$$\tilde{x} = \frac{20 - 10}{60 - 10} = 0.20.$$

So the triangular distribution is T(10, 20, 60) and the associated standard triangular distribution is T(0, 1, 0.20).

Beta

Recall the beta distributionhas two parameters (k_1, k_2) where $k_1 > 0$ and $k_2 > 0$, and takes on many shapes depending on the values of the parameters. The variable denoted as x, lies within two limits(a and b) where $(a \le x \le b)$. The distribution takes on many shapes, where it can skew to the right, skew to the left, where the mode is at either of the limit end points (a, b), includes various bathtub configurations, and also has symmetrical and uniform shapes. These shapes depend on the values of the two parameters, k_1, k_2. Perhaps the most common situations occur when $k_2 > k_1 > 1$ whereby the mode is greater than the low limit, a, and the distribution is skewed to the right. This is the distribution of interest in this chapter.

When the analyst wants to use this distribution in a simulation model and has no empirical or sample data to estimate the four parameters, he/she may turn to one or more experts who could provide estimates of the following type:

$\hat{a}$ = an estimate of the minimum value of x

$\hat{b}$ = an estimate of the maximum value of x

$\hat{\mu}$ = an estimate of the mean of x

$\hat{\tilde{x}}$ = an estimate of the most likely value of x

Recall, for the beta distribution, the mean and mode of x are computed from the parameters as follows:

$$\mu = \hat{a} + [k_1/(k_1 + k_2)](\hat{b} - \hat{a})$$

$$\tilde{x} = \hat{a} + [(k_1 - 1)/(k_1 + k_2 - 2)](\hat{b} - \hat{a})$$

Note there are two equations and two unknowns (k_1, k_2) when estimates of $(\hat{a}, \hat{b}, \hat{\mu}, \hat{\tilde{x}})$ are given. Using algebra, and the estimates provided, it is possible to estimate the unknown shape parameters, k_1, k_2 with the two equations listed below:

$$\hat{k}_1 = \left[(\hat{\mu} - \hat{a})(2\hat{\tilde{x}} - \hat{a} - \hat{b})\right] / \left[(\hat{\tilde{x}} - \hat{\mu})(\hat{b} - \hat{a})\right]$$

$$\hat{k}_2 = \left[(\hat{b} - \hat{\mu})\hat{k}_1\right] / \left[(\hat{\mu} - \hat{a})\right]$$

So now the analyst can use the beta distribution with estimates of all four parameters, $(\hat{k}_1, \hat{k}_2, \hat{a}, \hat{b})$ in the simulation model.

Example 11.4 Assume a simulation model where the analyst wants to use the beta distribution but has no empirical or sample data to estimate the parameters. The analyst gets advice from an expert(s) that provides estimates of $\hat{a} = 10$, $\hat{b} = 60$, $\hat{\mu} = 30$ and $\hat{\tilde{x}} = 20$. Using the above equations, the estimates of the parameters become,

$$\hat{k}_1 = [(30 - 10)(2 \times 20 - 10 - 60)]/[(20 - 30)(60 - 10)] = 1.2$$

$$\hat{k}_2 = [(60 - 30)\ 1.2]/(30 - 10) = 1.8$$

Hence, the beta distribution can now be applied with the parameters, $\hat{a} = 10$, $\hat{b} = 60$, $\hat{k}_1 = 1.2$, and $\hat{k}_2 = 1.8$.

Lognormal

Suppose a variable $x \geq \gamma$ where γ is a location parameter to x. Now let $x' = x - \gamma$ where $x' \geq 0$ and x' is lognormal. The corresponding normal variable to x' is $y = \ln(x')$ where ln is the natural logarithm, and thereby $x' = e^y$. The mean and variance of y are denoted as μ and σ^2, respectively, whereby $y \sim N(\mu, \sigma^2)$ and $x' \sim LN(\mu, \sigma^2)$. Note also $x = \gamma + x' = \gamma + e^y$. Assume the simulation analyst wants to apply the lognormal variable x in the simulation model but does not have any empirical or sample data to estimate the parameters. Instead, the analyst relies on expert(s) who are able to give the following estimates on the variable x:

γ = an estimate of the location parameter of x

$\tilde{x}$ = an estimate of the most likely value (mode) of x

x_α = an estimate of the α-quantile value of x

Note, the mode of x' and x are the following:

$$\tilde{x}' = e^{\mu - \sigma^2}$$

$$\tilde{x} = \gamma + e^{\mu - \sigma}$$

respectively. The α-quantile value of x becomes,

$$x_\alpha = \gamma + e^{\mu + z_\alpha \sigma}$$

where $z \sim N(0,1)$ and $P(z \leq z_\alpha) = \alpha$.
   Now note,

$$\mu - \sigma^2 = \ln(\tilde{x} - \gamma)$$

$$\mu + z_\alpha \sigma = \ln(x_\alpha - \gamma)$$

Applying some algebra,

$$[-\sigma^2 - z_\alpha \sigma] = \ln[(\tilde{x} - \gamma)/(x_\alpha - \gamma)] = c$$

Solving for σ via the quadratic equation,

$$\hat{\sigma} = \left[-z_\alpha + \sqrt{z_\alpha^2 - 4c} \right]/2$$

The estimate of the mean of y is below.

$$\hat{\mu} = \hat{\sigma}^2 + \ln(\tilde{x} - \gamma)$$

Finally, the analyst can now apply the lognormal distribution to the variable x using the parameters:

γ = location parameter of x

$\hat{\mu}$ = mean of y

$\hat{\sigma}$ = standard deviation of y

In essence, $x \sim LN(\gamma, \hat{\mu}, \hat{\sigma}^2)$.

Example 11.5 A simulation model is being developed and the analyst wants to use the lognormal distribution but has no empirical or sample data to estimate the parameters. The analyst gets advice from an expert(s) who provides estimates of $\hat{\gamma} = 100$, $\tilde{x} = 200$ and $x_{0.9} = 800$. Note $\alpha = 0.90$ and $z_{0.90} = 1.282$. Using the above results, the estimates of the parameters become,

$$c = \ln[(200 - 100)/(800 - 100)] = -1.946$$

$$\hat{\sigma} = \left[-1.282 + \sqrt{(1.282)^2 - 4(-1.946)} \right]/2 = 0.894$$

$$\hat{\mu} = 0.894^2 + \ln(200 - 100) = 5.404$$

Hence, the lognormal distribution can now be applied in the simulation model with the parameters, $\hat{\gamma} = 100$, $\hat{\mu} = 5.404$ and $\hat{\sigma} = 0.894$.

A quick check to ensure the estimates are correct is to measure the mode and/or the α-quantile that were provided at the outset. The computations are below.

$$\tilde{x} = \gamma + e^{\mu - \sigma^2} = 100 + e^{(5.404 - 0.894^2)} = 200$$

$$x_\alpha = \gamma + e^{\mu + z_\alpha \sigma} = 100 + e^{(5.404 + 1.282 \times 0.894)} = 800$$

Since the measures are the same as the specifications provided, ($\tilde{x} = 200$ and $x_{0.90} = 800$),the computations are accepted.

Weibull

Suppose a variable $x \geq \gamma$ where γ is a location parameter to x. Now let $x' = x - \gamma$ where $x' \geq 0$ and x' is Weibull distributed with parameters (k_1, k_2). Assume the simulation analyst wants to apply the Weibull distribution in the simulation system but does not have any empirical data to estimate the parameters. Instead, the analyst relies on expert(s) who are able to give the following estimates on the variable x:

γ = an estimate of the location parameter of x

$\tilde{x}$ = an estimate of the most likely value (mode) of x

x_α = an estimate of the α-quantile value of x

When $k_1 < 1$, the mode of x' is at x' = 0. The analysis here is when $k_1 \geq 1$ and the mode of x' is greater than zero. For this situation, the mode is measured as below.

$$\tilde{x}' = k_2[(k_1 - 1)/k_1]^{1/k1}$$

The corresponding mode of x is

$$\tilde{x} = \gamma + k_2[(k_1 - 1)/k_1]^{1/k1}$$

Using algebra, k_2 becomes

$$k_2 = (\tilde{x} - \gamma)/[(k_1 - 1)/k_1]^{1/k1}$$

The cumulative distribution for the α-quantile is obtained by the following,

$$F(x_\alpha) = 1 - \exp\left\{-[(x_\alpha - \gamma)/k_2]^{k1}\right\} = \alpha$$

Hence,

$$\ln(1 - \alpha) = -[(x_\alpha - \gamma)/k_2]^{k1}$$

Applying algebra and solving for k_2,

$$k_2 = (x_\alpha - \gamma)/\{\ln[1/(1 - \alpha)]\}^{1/k1}$$

So now,

$$(\tilde{x} - \gamma)/[(k_1 - 1)/k_1]^{1/k_1} = (x_\alpha - \gamma)/\ln[1/(1 - \alpha)]^{1/k_1}$$

whereby,

$$(\tilde{x} - \gamma)/(x_\alpha - \gamma) = \{(k_1 - 1)/[k_1 \times \ln[1/(1 - \alpha)]]\}^{1/k_1}$$

Solving for k_1

Because estimates of $\tilde{x}$, γ and x_α are provided, along with α, the only unknown in the above equation is k_1. At this point, an iterative search is made to find the value of k_1 where the right-hand-side of the above equation is equal to the left-hand-side. The result is $\hat{k}_1$.

Solving for k_2

Having found $\hat{k}_1$, the other parameter, k_2, is now obtained from

$$\hat{k}_2 = (\tilde{x} - \gamma)/[(\hat{k}_1 - 1)/\hat{k}_1]^{1/\hat{k}_1}$$

Example 11.6 A simulation model is being developed and the analyst wants to use the Weibull distribution but has no empirical or sample data to estimate the parameters. The analyst gets advice from an expert(s) that provides estimates of $\hat{\gamma} = 100$, $\tilde{x} = 130$ and $x_{0.9} = 500$. Note $\alpha = 0.90$. To find the estimate of k_1, the following computations are needed to begin the iterative search:

$$(\tilde{x} - \gamma)/(x_\alpha - \gamma) = (130 - 100)/(500 - 100) = 0.075$$

$$\{(k_1 - 1)/[k_1 \times \ln[1/(1 - \alpha)]]\}^{1/k_1} = \{(k_1 - 1)/[k_1 \times \ln[1/(1 - 0.9)]]\}^{1/k_1}$$
$$= \{(k_1 - 1)/[k_1 \times 2.302]\}^{1/k_1}$$

Note the left-hand-side (LHS) of the equation below. An iterative search of k_1 is now followed until the LHS is near to 0.075.

$$\text{LHS} = \{(k_1 - 1)/[k_1 \times 2.302]\}^{1/k_1} = 0.075$$

The search for k_1 begins with $k_1 = 2.00$, and continues until $k_1 = 1.14$:

At $k_1 = 2.00$, LHS $= 0.46$
At $k_1 = 1.50$, LHS $= 0.26$
At $k_1 = 1.20$, LHS $= 0.11$
At $k_1 = 1.14$, LHS $= 0.075$

Hence, $\hat{k}_1 = 1.14$.
So now, the estimate of k_2 is the following:

$$\hat{k}_2 = (\tilde{x} - \gamma)/[(\hat{k}_1 - 1)/\hat{k}_1]^{1/\hat{k}_1}$$
$$= (130 - 100)/[(1.14 - 1.00)/1.14]^{1/1.14}$$
$$= 188.9$$

Finally, the estimates of the parameters are ($\hat{k}_1 = 1.14$, $\hat{k}_2 = 188.9$).

A quick check to ensure the estimates are correct, requires measuring the mode and/or the α-quantile and compare the measures with those that were provided at the outset. The computations are below.

$$\tilde{x} = \gamma + k_2[(k_1 - 1)/k_1]^{1/k_1}$$
$$= 100 + 188.9[(1.14 - 1.00)/1.14]^{1/1.14}$$
$$= 130$$

$$x_\alpha = \gamma + k_2\{\ln[1/(1 - \alpha)]\}^{1/k1}$$
$$= 100 + 188.9 \ \ln[1/(1 - 0.90]^{1/1.14}$$
$$= 492.5$$

Since the above measures are sufficiently near the data provided, ($\tilde{x} = 130$ and $x_{0.90} = 500$) the parameters estimated are accepted.

Summary

Sometimes the analyst may need to develop a computer simulation model that includes one or more variables where no empirical or sample data is available. This is where he/she seeks opinions from one or more experts who give some estimates on the characteristics of the variable. The chapter pertains to these situations and shows some of the common ways to select the probability distribution and estimate the associated parameters.

Appendix A

Table A.1 Measures from the standard normal distribution

F(z)	z	F(z)	z
0.010	2.327	0.510	0.025
0.020	−2.054	0.520	0.050
0.030	−1.881	0.530	0.075
0.040	−1.751	0.540	0.100
0.050	−1.645	0.550	0.125
0.060	−1.555	0.560	0.151
0.070	−1.476	0.570	0.176
0.080	−1.405	0.580	0.202
0.090	−1.341	0.590	0.227
0.100	−1.282	0.600	0.253
0.110	−1.227	0.610	0.279
0.120	−1.175	0.620	0.305
0.130	−1.126	0.630	0.331
0.140	−1.080	0.640	0.358
0.150	−1.036	0.650	0.385
0.160	−0.994	0.660	0.412
0.170	−0.954	0.670	0.439
0.180	−0.915	0.680	0.467
0.190	−0.878	0.690	0.495
0.200	−0.841	0.700	0.524
0.210	−0.806	0.710	0.553
0.220	−0.772	0.720	0.582
0.230	−0.739	0.730	0.612
0.240	−0.706	0.740	0.643
0.250	−0.674	0.750	0.674
0.260	−0.643	0.760	0.706
0.270	−0.612	0.770	0.739
0.280	−0.582	0.780	0.772
0.290	−0.553	0.790	0.806
0.300	−0.524	0.800	0.841

(continued)

N.T. Thomopoulos, *Essentials of Monte Carlo Simulation: Statistical Methods for Building Simulation Models*, DOI 10.1007/978-1-4614-6022-0, © Springer Science+Business Media New York 2013

Table A.1 (continued)

F(z)	z	F(z)	z
0.310	−0.495	0.810	0.878
0.320	−0.467	0.820	0.915
0.330	−0.439	0.830	0.954
0.340	−0.412	0.840	0.994
0.350	−0.385	0.850	1.036
0.360	−0.358	0.860	1.080
0.370	−0.331	0.870	1.126
0.380	−0.305	0.880	1.175
0.390	−0.279	0.890	1.227
0.400	−0.253	0.900	1.282
0.410	−0.227	0.910	1.341
0.420	−0.202	0.920	1.405
0.430	−0.176	0.930	1.476
0.440	−0.151	0.940	1.555
0.450	−0.125	0.950	1.645
0.460	−0.100	0.960	1.751
0.470	−0.075	0.970	1.881
0.480	−0.050	0.980	2.054
0.490	−0.025	0.990	2.327
0.500	−0.000		

Table A.2 Probability distributions, random variables, notation and parameters	Continuous distributions:	
	Standard uniform	$u \sim U(0,1)$
	Continuous uniform	$x \sim CU(a,b)$
	Exponential	$x \sim Exp(\theta)$
	Erlang	$x \sim Erl(k, \theta)$
	Gamma	$x \sim Gam(k, \theta)$
	Beta	$x \sim Beta(k_1,k_2,a,b)$
	Weibull	$x \sim We(k_1,k_2,\gamma)$
	Normal	$x \sim N(\mu,\sigma^2)$
	Lognormal	$x \sim LN(\mu_y,\sigma_y^2)$
	Triangular	$x \sim TR(a,b,\hat{x}))$
	Discrete distributions:	
	Discrete uniform	$x \sim DU(a,b)$
	Bernoulli	$x \sim Be(p)$
	Binomial	$x \sim Bin(n,p)$
	Geometric	$x \sim Ge(p)$
	Pascal	$x \sim Pa(k,p)$
	Hyper geometric	$x \sim HG(n,N,D)$
	Poisson	$x \sim Po(\theta)$
	Multivariate distributions:	
	Multivariate arbitrary	$x_1,\ldots,x_k \sim MA(p_{1\ldots k})$
	Multinomial	$x_1,\ldots,x_k \sim MN(n,p_1,\ldots,p_k)$
	Multivariate hyper geometric	$x_1,\ldots,x_k \sim MHG(n,N,D_1,\ldots,D_k)$
	Bivariate normal	$x_1, x_2 \sim BVN(\mu_1,\mu_2,\sigma_1,\sigma_2,\rho)$
	Bivariate lognormal	$x_1, x_2 \sim BVLN(\mu_{y1},\mu_{y2},\sigma_{y1},\sigma_{y2},\rho_y)$
	Multivariate normal	$x_1,\ldots,x_k \sim MVN(\mu,\Sigma)$
	Multivariate lognormal	$x_1,\ldots,x_k \sim MVLN(\mu_y,\Sigma_y)$

Table A.3 Continuous uniform u ~ U(0,1) random variates

0.3650	0.4899	0.1557	0.4745	0.2573	0.6288	0.5421	0.1563
0.5061	0.3905	0.1074	0.7840	0.4596	0.7537	0.5961	0.8327
0.0740	0.1055	0.3317	0.1282	0.0002	0.5368	0.6571	0.5440
0.1919	0.6789	0.4542	0.3570	0.1500	0.7044	0.9288	0.5302
0.4018	0.4619	0.4922	0.2076	0.3297	0.0954	0.5898	0.1699
0.4439	0.2729	0.8725	0.7507	0.2729	0.6736	0.2566	0.0899
0.7901	0.2973	0.2353	0.4805	0.2546	0.3406	0.0449	0.4824
0.5886	0.7549	0.9279	0.3310	0.5429	0.0807	0.6344	0.4100
0.9234	0.6202	0.3477	0.1492	0.4800	0.2194	0.9937	0.1304
0.5477	0.9230	0.5382	0.4064	0.8472	0.8262	0.6724	0.7219
0.4952	0.4130	0.6953	0.1791	0.4229	0.5432	0.8147	0.5409
0.2278	0.6192	0.4898	0.6808	0.8866	0.3705	0.3025	0.2929
0.2233	0.5845	0.3635	0.8760	0.4780	0.1906	0.6841	0.7474
0.1617	0.8078	0.2026	0.9568	0.0659	0.0615	0.7932	0.3796
0.1155	0.1738	0.0481	0.7148	0.5330	0.5610	0.2167	0.4680
0.3989	0.9031	0.7460	0.0886	0.6346	0.7130	0.0157	0.4311
0.9854	0.8026	0.6961	0.4176	0.7345	0.2772	0.3566	0.4335
0.6460	0.3478	0.1044	0.1854	0.0777	0.4328	0.9593	0.5420
0.2178	0.3790	0.3958	0.2815	0.5034	0.1387	0.5173	0.9654
0.6573	0.4411	0.6930	0.0645	0.7561	0.7005	0.4971	0.1554
0.7845	0.0503	0.5180	0.7570	0.8007	0.3252	0.9727	0.8043
0.8758	0.4166	0.1231	0.9542	0.7973	0.6963	0.4016	0.0163
0.5097	0.4061	0.1061	0.2761	0.6430	0.8491	0.4980	0.1878
0.3236	0.7708	0.2180	0.4470	0.2360	0.8784	0.6104	0.3744
0.5859	0.9316	0.5172	0.3303	0.8685	0.2591	0.2595	0.1787
0.7423	0.8409	0.2786	0.7030	0.4049	0.8116	0.7418	0.4377
0.3394	0.7106	0.3123	0.7988	0.1518	0.5930	0.9562	0.2431
0.9843	0.6330	0.5989	0.9026	0.5749	0.2452	0.8602	0.0750
0.2457	0.3786	0.3972	0.5266	0.2704	0.5812	0.2097	0.0787
0.6524	0.9003	0.2316	0.9499	0.8462	0.4412	0.4920	0.7695
0.1955	0.3262	0.4132	0.1527	0.6198	0.0994	0.2050	0.6925
0.9914	0.4714	0.0040	0.4258	0.2887	0.7525	0.8913	0.8219
0.0103	0.1517	0.3774	0.1881	0.9795	0.8721	0.5815	0.7294
0.0282	0.8279	0.7834	0.7912	0.3327	0.4509	0.5551	0.8033
0.2076	0.3647	0.5735	0.3442	0.5282	0.4255	0.5730	0.0500
0.9627	0.9331	0.9926	0.8396	0.4093	0.8053	0.9894	0.2584
0.0170	0.3391	0.6925	0.1104	0.1097	0.2906	0.3989	0.5590
0.8079	0.3096	0.3758	0.4010	0.8414	0.4096	0.7246	0.6588
0.6456	0.5161	0.2233	0.5828	0.7485	0.4565	0.9044	0.2830
0.2814	0.3681	0.0142	0.2947	0.9840	0.7613	0.5809	0.6057

Table A.4 Standard normal random variates z ~ N(0,1)

−0.058	1.167	0.948	−0.173	−1.114	−0.823	1.163	−0.847
−1.843	−0.465	−0.503	−2.375	0.416	−0.291	1.131	0.533
−0.369	0.207	−1.338	0.367	−0.019	1.885	−1.382	0.321
0.776	0.003	0.168	−0.817	−0.380	−0.852	−0.026	1.273
0.567	0.303	0.302	−2.234	−0.068	1.506	−0.891	−0.292
−0.460	1.049	−0.086	−0.627	−0.922	1.526	−0.500	−1.494
−1.204	−1.251	−0.585	0.822	1.785	−0.661	−0.817	1.110
−1.005	−1.529	−0.219	−0.506	−0.662	0.638	−1.243	0.528
−0.636	1.845	2.548	1.332	0.587	−0.320	0.385	−0.674
−1.311	0.065	−0.183	0.841	−1.055	0.282	−1.208	−0.364
0.320	−0.114	−0.752	0.091	−0.614	−0.174	−0.736	1.151
1.169	−1.373	0.067	−0.288	−0.553	−0.746	1.651	0.127
2.710	−1.830	0.061	−0.102	1.228	1.074	−1.635	0.383
−0.019	−0.044	0.580	−0.596	2.391	−1.648	0.382	−0.701
1.580	0.992	−0.465	−0.452	−0.840	−2.280	0.237	−0.182
−1.329	1.847	−0.599	0.213	1.323	−0.629	0.030	0.447
−1.022	1.652	−1.785	0.840	−0.771	1.062	0.425	0.253
−0.212	0.098	1.578	−1.564	2.791	−0.890	−1.356	1.868
−1.049	−0.556	−0.350	1.569	0.482	−0.604	−0.524	0.486
−0.122	−2.494	−0.842	−0.630	1.341	0.364	1.270	−0.139
0.545	−1.334	−0.614	1.533	0.966	0.020	0.938	0.312
−0.512	0.867	1.187	0.313	−0.480	−0.069	−0.045	0.720
−0.193	−0.386	−0.030	−0.472	−1.273	0.230	0.357	−0.471
−0.836	−1.022	−0.288	2.560	−0.125	−1.392	−0.255	−1.256
1.784	0.587	−1.051	−0.648	1.813	0.322	0.280	−1.066
−0.547	1.636	−0.219	0.409	1.953	1.191	0.688	1.230
−0.477	−0.120	0.869	−0.199	0.270	1.595	−0.745	0.324
−1.096	−0.362	−1.561	−0.843	0.301	0.478	−1.170	0.473
2.071	0.791	1.278	0.672	1.145	−1.655	−0.173	−0.505
0.061	1.781	−0.265	1.101	−1.535	2.265	0.219	0.771
−0.526	−0.385	−0.278	0.762	−0.514	−0.132	−0.456	0.244
−0.527	−0.138	1.715	−1.463	−1.007	−1.651	−0.099	1.421
−1.220	0.651	−1.251	−1.132	1.338	0.462	−2.048	0.369
−0.315	0.677	0.425	−1.238	1.432	0.527	−0.077	2.720
1.036	0.195	−0.095	0.787	−0.251	−0.577	−0.401	0.666
−0.754	1.024	−1.087	0.073	−0.672	−1.405	3.332	0.964
0.288	−0.456	−1.264	−0.685	−0.234	0.049	0.032	−1.068
−1.299	0.699	−0.775	−0.232	−1.773	0.352	1.175	0.451
−0.417	0.995	0.791	−1.750	1.436	−1.364	0.797	−0.036
−1.402	0.500	0.409	−0.858	−0.322	−0.407	−1.502	−0.523
−0.031	−2.155	0.615	−0.612	−1.195	−0.519	−1.559	1.558
0.483	0.223	1.511	0.493	−0.773	0.116	0.349	−1.661

Table A.5 Standard exponential random variates when E(x) = 1.00

2.247	2.012	0.485	3.367	0.699	4.144	0.592	1.51
2.118	0.866	1.978	1.659	1.46	2.577	0.052	1.742
0.054	2.334	4.419	0.304	0.776	0.42	0.24	0.225
0.882	1.945	0.092	0.224	1.005	0.617	0.337	1.793
0.506	1.569	1.639	0.936	0.44	0.881	1.081	0.875
0.061	0.164	0.38	0.572	0.551	1.941	0.058	2.276
1.516	0.037	0.147	1.106	0.305	1.847	0.236	0.256
0.933	0.106	2.865	0.482	0.427	0.041	2.209	0.68
0.289	0.808	0.045	1.683	0.635	0.789	0.473	0.159
0.236	0.39	0.551	1.352	0.126	0.572	1.741	0.852
0.073	0.978	0.698	1.166	0.434	1.002	0.175	2.371
0.179	0.691	0.034	1.377	1.837	0.349	0.547	0.722
3.754	0.098	1.516	0.045	3.378	8.186	1.167	1.164
3.402	1.187	2.164	0.136	0.163	0.315	0.032	0.392
1.207	1.777	0.091	1.706	1.674	0.226	0.11	0.978
0.619	2.865	0.214	0.354	0.933	0.283	0.375	0.098
1.323	0.558	0.615	0.617	1.94	0.53	1.227	0.472
0.258	0.072	1.562	1.729	0.404	0.545	5.107	1.201
0.415	0.165	2.362	0.715	1.553	0.954	0.424	3.283
0.247	1.487	2.243	0.121	2.342	0.972	2.208	1.414
1.738	0.166	1.576	0.56	0.785	0.074	0.058	1.804
1.069	3.084	0.128	0.485	0.679	2.303	0.186	1.745
0.048	2.424	0.675	1.11	0.068	0.014	3.12	0.236
0.186	0.339	1.387	1.923	0.296	0.427	2.025	1.956
0.21	2.888	0.299	0.018	0.389	0.917	0.476	1.028
0.204	0.059	0.282	0.221	3.73	0.195	1.067	2.397
1.474	1.421	0.493	0.169	0.846	0.128	0.317	1.945
0.835	0.264	0.016	0.295	3.046	0.057	1.916	0.084
2.436	0.055	0.912	0.692	2.082	0.589	1.776	1.751
0.709	1.572	0.563	3.776	0.153	1.117	1.18	1.322
1.918	0.19	1.091	0.323	2.044	1.323	0.502	0.103
1.119	0.371	0.091	1.002	0.796	0.866	1.837	2.713
1.259	1.665	0.386	1.891	1.505	0.482	0.878	0.463
0.884	1.49	0.395	0.305	2.009	0.3	0.173	0.175
0.742	3.262	0.501	0.353	1.115	0.263	1.18	0.398
0.319	1.123	6.492	0.478	0.455	0.612	1.27	0.295
0.624	0.134	0.222	0.281	5.387	0.016	0.397	1.244
0.161	0.39	0.648	2.583	0.04	1.441	0.077	0.994
0.519	1.427	0.191	2.758	0.392	0.072	0.356	1.05
0.178	0.571	2.612	0.629	4.506	0.565	1.453	1.249

Appendix B

Problems

In solving the problems, the student will occasionally need one or more random variates of the uniform type, $u \sim U(0,1)$, or of the standard normal type, $z \sim N(0,1)$. These are provided in the Appendix with random variates of each type, Table A.3 for $u \sim U(0,1)$ and Table A.4 for $z \sim (N(0,1)$. On each problem, the student should begin on the first row and first column to retrieve the random variate, then the first row and second column, and so forth. Hence, for $u \sim U(0,1)$, the variates are: 0.3650, 0.4899, and so forth. For $z \sim N(0,1)$, they are: -0.058, 1.167, so forth.

Chapter 2

2.1 Using the Linear Congruent method with parameters, $m = 16$, $a = 5$, $b = 1$ and the seed $w_0 = 7$, list the next 16 entries of w_i for $i = 1$ to 16.

2.2 Use the Linear Congruent method with parameters, $a = 20$, $b = 0$, $m = 64$ and $w_0 = 3$ to generate the next three entries of w.

2.3 Use the results of Problem 2.2 and list the three entries of the corresponding uniform, u, distribution

Chapter 3

3.1 The variable x is continuous with probability density $f(x) = 3/8x^2$ for $0 \leq x \leq 2$. Use the inverse transform method to generate a random variate of x.

N.T. Thomopoulos, *Essentials of Monte Carlo Simulation: Statistical Methods for Building Simulation Models*, DOI 10.1007/978-1-4614-6022-0,
© Springer Science+Business Media New York 2013

3.2   The variable x is discrete with probability distribution:

x	−1	0	1	2	3	4
p(x)	0.10	0.20	0.30	0.20	0.15	0.05

Use the inverse transform method to generate a random variate of x.

3.3   The variable x is continuous with probability density $f(x) = 3/8x^2$ for $0 \leq x \leq 2$. Use the Accept-Reject method to generate a random variate of x.

3.4   The variable x is continuous with probability density $f(x) = 3/8x^2$ for $0 \leq x \leq 2$. Use the inverse transform method to generate a random variate of x that is restricted to lie within 1 and 2.

3.5   The variable x is continuous with probability density $f(x) = 3/8x^2$ for $0 \leq x \leq 2$. Use the inverse transform method to generate a random variate of $y = \min(x_1, \dots, x_8)$.

3.6   The variable x is continuous with probability density $f(x) = 3/8x^2$ for $0 \leq x \leq 2$. Use the inverse transform method to generate a random variate of $y = \max(x_1, \dots, x_8)$.

3.7   The variable x is composed of three distributions, $f_1(x)$, $f_2(x)$ and $f_3(x)$ with probabilities 0.5, 0.3, 0.2, respectively. $f_1(x) = 0.1$ for $0 < x < 10$, $f_2(x) = 0.2x$ for $0 < x < 10$ and $f_3(x) = 0.003x^2$ for $0 < x, <10$. Generate one random variate of x.

3.8   The variable x is continuous with density $f(x) = 0.25x^3$ $0 < x < 2$. Generate a random variate of $y = x_1 + x_2 + x_3$.

3.9   The variable x is triangular with a minimum value of -10, mode of 0 and maximum of 20. Generate a random variate of x.

3.10 The variable x is continuous with empirical data of: 5. 8, 12, 20, 25, 3, 6, 10,15. Generate one random variate of x using the composition method.

3.11 The variable x is continuous with grouped empirical data as follows:

[a,b)	Frequency
[0, 10)	36
[10–20)	10
[20–30)	4

Generate one random variate of x using the composition method.

Chapter 4

4.1   The variable x is a continuous uniform for $-10 < x < 30$. Use the inverse transform method to generate a random variate of x.

4.2   The variable x is exponential with $E(x) = 5$. Generate a random variate of x.

4.3   The variable x is Erlang with $k = 5$ and $E(x) = 20$. Generate a random variate of x.

4.4 The variable x is Gamma with sample data of $\bar{x} = 5.0$ and $s^2 = 10.0$. Generate a random variate of x.

4.5 The variable x is Gamma with sample data of $\bar{x} = 1.0$ and $s^2 = 10.0$. Generate a random variate of x.

4.6 The variable x is Beta with parameters, $(k_1, k_2) = (1, 8)$ and $(a, b) = (10, 90)$. Two random Gammas are $g_1 = 13$ for $(k_1, k_2) = (10,1)$ and $g_2 = 20$ for $(k_1, k_2) = (8,1)$. Find the random variate for the Beta.

4.7 The variable x is Weibull with parameters, $(k_1, k_2) = (2, 20)$. Generate a random variate of x.

4.8 The variable x is Normal with mean $= 100$ and variance $= 100$. Use the Sine-Cosine method to generate a random variate of x.

4.9 The variable x is Lognormal with $\mu_x = 5$ and $\sigma_x = 100$. Generate a random variate of x.

4.10 Generate a random variate for a chi-square variable with degrees of freedom $= 5$.

4.11 Generate a random variate for a chi-square variable with degrees of freedom $= 153$.

4.12 Generate a random variate for a student's t with degrees of freedom $= 5$. Get z first.

4.13 Generate a random variate for an F distribution with degrees of freedom 5 and 2. Get chi square $(df = 5)$, first.

Chapter 5

5.1 The variable x is discrete with the following probability distribution.

x	−1	0	1	2
P(x)	0.50	0.30	0.15	0.05

Generate one random variate of x.

5.2 The variable x is from a discrete uniform distribution where x ranges from −5 to 5. Generate a random variate of x.

5.3 Generate a random variate of x for a Bernoulli with $p = 0.30$.

5.4 Generate a random variate of x for a Binomial with $n = 10$ and $p = 0.30$.

5.5 Generate a random variate of x for a Binomial with $n = 500$ and $p = 0.20$.

5.6 Generate a random variate of x for a Binomial with $n = 500$ and $p = 0.001$.

5.7 Generate a random variate of x for a Hyper Geometric with $N = 20$, $D = 5$ and $n = 2$.

5.8 Generate a random variate of x for a Geometric with $p = 0.30$. Note, x is the number of trials till the first success.

5.9 Generate a random variate for a Pascal with $p = 0.30$ and $k = 5$. Note, x is the number of trials till five successes.

5.10 Generate a random variate for a Poisson with $E(x) = 1.5$.

Chapter 6

6.1 The variables x_1, x_2, x_3, x_4 are jointly related by the following probabilities:

		x_1	0		1	
		x_2	1	2	1	2
x_3	x_4					
0	3		.01	.09	.02	.10
	4		.03	.01	.04	.02
1	3		.05	.06	.06	.09
	4		.07	.12	.08	.15

Generate one set of random variates for x_1, x_2, x_3, x_4.

6.3 Consider the multinomial distribution with $k = 4$, $p_1 = .4$, $p_2 = .3$, $p_3 = .2$, $p_4 = .1$ and $n = 5$. Generate one set of random variates for x_1, x_2, x_3, x_4.

6.4 Consider x_1, x_2 that are related by the bivariate normal distribution with $\mu_1 = 1.0$, $\mu_2 = 0.8$, $\sigma_1 = 0.2$, $\sigma_2 = 0.1$ and $\rho = 0.5$. Generate one random variate set of x_1, x_2.

6.6 Consider x_1, x_2 that are related by the bivariate lognormal distribution with parameters $\mu_{y1} = 20$, $\mu_{y2} = 2$, $\sigma_{y1} = 2$, $\sigma_{y2} = 0.5$ and $\rho_y = 0.8$, where $y_i = \ln(x_i)$ for $i = 1,2$. Generate one random variate set of x_1, x_2.

6.8 Generate the Cholesky matrix from the variance-covariance matrix:

$$\begin{bmatrix} 16 & 4 & 2 \\ 4 & 4 & 1 \\ 2 & 1 & 1 \end{bmatrix}$$

6.9 The variables x_1, x_2, x_3 are from a multivariate normal distribution with the following matrices. Generate one random set of x_1, x_2, x_3:

$$\mu = \begin{bmatrix} 20 \\ 6 \\ 10 \end{bmatrix} \quad C = \begin{bmatrix} 4 & 0 & 0 \\ 1 & 3 & 0 \\ 0.5 & 0.5 & 0.5 \end{bmatrix}$$

6.10 The variables x_1, x_2, x_3 are from a multivariate lognormal distribution with the following matrices from the transformed values of y_1, y_2, y_3. Generate one random set of x_1, x_2, x_3:

$$\mu_y = \begin{bmatrix} 2.0 \\ 0.6 \\ 1.0 \end{bmatrix} \quad C_y = \begin{bmatrix} 4 & 0 & 0 \\ 1 & 3 & 0 \\ 0.5 & 0.5 & 0.5 \end{bmatrix}$$

Chapter 7

7.1 Consider a Poisson process where A(j) = arrival rate, B(j) = time for j = 1, 2, 3, 4:

j	1	2	3	4	. . . . . .
A(j)	2	3	5	4	
B(j)	0	1	2	3	

Generate the random times of the first three arrivals.

7.2 Generate the batch size, x, for an arrival where x = y + 1 and y is Posson distributed with E(y) = 2.4.

7.3 Generate a random variate, x, that is the time to fail for a unit that has four active redundant units, with times denoted as y, that are Exponential and E(y) = 100.

7.5 Generate a random variate, x, that is the time to fail for a unit that has four standby redundant units, with times denoted as y, that are Exponential and E(y) = 100.

7.7 From the integers of 1–20, generate the sequence of n = 5 of them randomly without replacement.

7.8 From a deck of N = 52 regular cards, generate a random hand for a player who will receive five of the cards. Use the same index to identify cards as listed in the text.

Chapter 9

9.2 Simulation results of n = 11 runs yields $\bar{x} = 100$ and s = 18. Compute the 0.95 confidence limits for the true mean. Note, $t_{10, 0.025} = 2.228$.

9.3 In Problem 9.2, the analyst wants (U – L) = 5.0. How many more runs are needed?

9.4 Simulation results of n = 40 runs yield w = 8 units with an attribute. Find the 0.95 confidence limits on the true proportion of units with the attribute.

9.5 In Problem 9.4, the analyst wants (U – L) = 0.04. How many more runs are needed?

9.6 Consider a machine shop where an order of No = 15 is needed. A simulation is run where the units started is Ns = 20, and after n = 1,000 simulation runs, Ng = 958 is the number of good units. The management wants to be 95% certain that the number of good units will exceed No. Find the 0.95 confidence interval on the probability that Ng will be equal or larger to No.

9.7 For Problem 9.6, how many more runs, if any, are needed so that the 0.95 confidence interval (from L to U) is always above the 0.95 specification mark?

9.9 Suppose two options are run in a simulation with results of $n_1 = 20$, $\bar{x}_1 = 100_1$ $= 100$ and $s_1 = 30$ for option 1, and $n_2 = 20$, $\bar{x}_2 = 95$ and $s_2 = 25$ for option 2. Assuming the two variances are the same, find the 0.95 confidence limits on the difference between the two means.

9.10 A simulation is run with four options (i) and four observations (j) of each are run with results in the table below. Use the one-way-analysis-of-variance method to determine if all the means of the four options are the same at the 0.05 significance level. Note $F_{3,12,0.05} = 3.49$.

	Observations (j)			
	1	2	3	4
Options (i)				
1	5	7	6	4
2	3	5	6	2
3	8	9	7	8
4	6	4	4	6

9.12 Using the results of Problem 9.10, compute the 0.95 confidence interval on the difference between each pair of means, and label any that are significantly different. Note, $t_{0.025,12} = 2.179$.

Chapter 10

10.2 A sample of $n = 10$ data entries are the following: (10, 13, 9, 7, 8, 12, 15, 10, 3, 8). Compute the following: $x(1)$, $x(n)$, $\bar{x}$, s, cov, τ.

10.3 From the data of Problem 10.2, compute the estimate of the location parameter, γ.

10.4 Consider the ($n = 15$) sample data from a continuous uniform distribution: (1.3, 1.4, 1.8, 2.3, 2.4, 2.5, 2.9, 3.1, 3.4, 3.9, 4.1, 4.7, 5.2, 5.7, 6.1). Find the maximum likelihood estimates of the min and max parameters, (a, b). Now find the method of moments estimators for a and b.

10.5 Suppose the $n = 12$ sample data entries are the following: (10.4, 12.3, 13.5, 14.6, 15.1, 15.8, 16.2, 16.5, 17.3, 16.3, 15.1, 19.4). Assuming the data are normally distributed, estimate the parameters of the mean and standard deviation.

10.6 From the $n = 8$ sample data: (0.7, 1.2, 1.8, 2.4, 4.0,10.3, 0.9, 1.4), estimate the parameter for an exponential distribution.

10.7 Assume the lognormal variable x, with sample data: (10, 12, 15, 23, 40, 90, 217). Estimate the parameters for this distribution.

10.8 Suppose the variable x is assumed as gamma distributed and a sample of $n = 50$ yields the sample average of 33 and the sample variance of 342. Estimate the parameters for the gamma distribution.

10.9 Suppose the variable x is assumed as beta distributed and estimates of the following are given: mean $= 60$, mode $= 80$, min $= 0$, max $= 100$. Estimate the parameters for the beta distribution.

10.10 The following ($n = 11$) sample data are: (3, 3, 5, 7, 8, 8, 11,12,13,15,16). Assuming the data comes from a discrete uniform distribution, estimate the maximum likelihood estimate for the min and max (a, b). Now estimate the parameters using the method of moments.

10.11 Suppose the variable x is from a binomial distribution with $n = 10$, p is unknown, and $m = 8$ samples of x are (3, 2, 1, 5, 3, 4, 3, 2). Find the maximum likelihood estimate of p.

10.12 Suppose the variable x is from a geometric distribution where p is unknown, and $m = 6$ samples of x are (3, 5, 8, 4, 7, 3). Find the maximum likelihood estimate of p. Recall, x $=$ number of failures till a success.

10.13 Suppose the variable x is from a Pascal distribution with $k = 3$, p is unknown, and $m = 10$ samples of x are (8, 9, 9, 12, 10, 13, 10, 12, 11, 13). Find the maximum likelihood estimate of p. Recall, x $=$ number of failures till three successes.

10.14 Suppose the variable x is from a Poisson distribution where the parameter is unknown, and $m = 7$ samples of x are (3, 5, 2, 7, 4, 4, 1). Find the maximum likelihood estimate of the parameter.

10.15 Consider the ($n = 15$) sample data from Problem 10.4 and the maximum likelihood estimates of the parameters. Assuming the continuous uniform distribution, list the vectors Xs and Xf for the Q-Q Plot.

10.16 Consider the ($n = 11$) sample data from Problem 10.10 and the maximum likelihood estimates of the parameters. Assuming the discrete uniform distribution pertains, list the vectors Fs and Ff for the P-P Plot.

Chapter 11

11.1 Suppose the variable x is from a continuous uniform distribution and an expert estimates the minimum value is 50 and the 0.75-quantile is 90. Estimate the parameters for this distribution.

11.2 Suppose the variable x is from a continuous uniform distribution and an expert estimates the maximum value is 100 and the 0.20-quantile is 40. Estimate the parameters for this distribution.

11.3 Assume a variable x from the triangular distribution where an expert estimates the following: min $= 5$, most likely $= 20$ and the max $= 30$. Estimate the parameters for the standard triangular distribution.

11.4 Assume a variable x from the beta distribution where an expert estimates the following: min $= 5$, mean $= 18$, most likely $= 20$ and the max $= 30$. Estimate the parameters for the beta distribution.

11.5 Assume a variable x from the lognormal distribution where an expert estimates the following: min = 0, most likely = 20 and the 0.95-quantile = 100. Estimate the parameters for the lognormal distribution.

11.6 Assume a variable x from the Weibull distribution where an expert estimates the following: min = 0, most likely = 20 and the 0.95-quantile = 100. Estimate the parameters for the Weibull distribution.

Appendix C

Solutions

2.1	4, 5, 10, 3. 0, 1, 6, 15, 12, 13, 2, 11, 8, 9, 14, 7
2.2	60, 48, 0
2,3	0.9375, 0.750, 0.000
3.1	1.429
3.2	1
3.3	1.084
3.4	1.526
3.5	1.260
3.6	1.918
3.7	4.899
3.8	4.484
3.9	0.500
3.10	6.978
3.11	4.899
4.1	4.600
4.2	2.271
4.3	8.948
4.4	4.230
4.5	0.00058
4.6	41.515
4.7	13.478.
4.8	114.18
4.9	1.283
4.10	3.535
4.11	151.98
4.12	0.063
4.13	0.697
5.1	−1
5.2	−1

(continued)

N.T. Thomopoulos, *Essentials of Monte Carlo Simulation: Statistical Methods for Building Simulation Models*, DOI 10.1007/978-1-4614-6022-0, © Springer Science+Business Media New York 2013

5.3	1			
5.4	3			
5.5	99			
5.6	1			
5.7	0			
5.8	2			
5.9	8			
5.10	3			
6.1	0,2,0,4			
6.3	3,0,2,0			
6.4	19.884, 2.326			
6.6	2.686, 2.454			
6.8	C =	4	0	0
		1	1.732	0
		0.5	0.289	0.666
6.9	X =	19.768		
		9.443		
		11.028		
7.1	0.908, 2.865, 3.562			
7.2	x = 6			
7.3	67.14			
7.5	193.76			
7.7	8, 11, 3, 12, 6			
7.8	8 H, AD, 6D, AS, KD			
9.2	(L,U) = (87.908, 112.091)			
9.3	188 more			
9.4	(L,U) = (0.076, 0.324)			
9.5	1,497 more'			
9.6	(L,U) = (0.946, 0.970)			
9.7	1,415 more			
9.8	(L,U) = (−12.11, 22.11)			
9.9	(L,U) = (−0.1917, 0.0317)			
9.10	F = 6.61 Significant			
9.12	1 v 2	0.476, 2.524 s		
	1 v 3	−3.524, −1.475 s		
	1 v 4	−0.524, 1.524 ns		
	2 v 3	−5.024, −2.976 s		
	2 v 4	−2.024, 0.024 ns		
	3 v 4	1.976, 4.024 s		
10.2	$x(1) = 3$			
	$x(10) = 15$			
	$\bar{x} = 9.5$			
	$s = 3.375$			
	$cov = 0.355$			
	$\tau = 1.190$			
10.3	−1			
10.4	MLE: a = 1.3, b = 6.1			
	MOM: a = 0.746, b = 6.025			

(continued)

10.5   N(15.208, 2.355^2)

10.6   0.352

10.7   LN(3.457, 1.139^2)

10.8 $\theta = 0.096$
       k = 3.168

10.9   (a, b, k$_1$, k$_2$) = (0,100,1.8,1.2)

10.10 MLE: a = 3, b = 16
       MOM: a = 1, b = 17

10.11 0.2875

10.12 0.167

10.13 0.219

10.14 $\theta = 3.714$

10.15 Xs = (1.3,1.4, 1.8, 2.3, 2.4, 2.5, 2.9, 3.1, 3.4, 3.9, 4.1,4.7, 5.2, 5.7, 6.1)
       Xf = (1.46, 1.78, 2.10, 2.42, 2.74, 3.06, 3.38, 3.70, 4.02, 4.34, 4.66, 4.98, 5.30, 5.62, 5.94)

10.16 Fs = (.045, .136, .227, .318, .409, 500, .591, .683, .774, .865, .955)
       Ff = (.071, .071, .214, .357, .429, .429, .643, .714, .786, .928, 1.000)

11.1   103.33

11.2   25

11.3   (0. 0.6, 1)

11.4   (5, 30, 1.3, 1.2)

11.5   LN(3.470, 0.686^2)

11.6   k$_1$ = 1.44
       k$_2$ = 45.54

References

Abramowitz, M., Stegun, I.A.: Handbook of Mathematical Functions with Formulas, Graphs and Tables, pp. 931–937. U.S. Department of Commerce, Washington (1964)

Ahrens, J.H., Dieter, V.: Computer methods for sampling from gamma, beta, poisson and binomial distributions. Computing **12**, 223–246 (1974)

Barnett, V.: Probability plotting methods and order statistics. Appl. Stat. **24**, 95–108 (1975)

Box, G.E.P., Miller, M.E.: A Note on the generation of random normal variates. Ann. Math. Stat. **29**, 610–611 (1958)

Cheng, R.C.H.: The Generation of gamma variables with non-integral shape parameter. Appl. Stat. **26**, 71–75 (1977)

Dubey, S.D.: On some permissible estimators of the location parameters of the Wiebull and certain other distributions. Technometrics **9**, 293–307 (1967)

Fishman, G.S., Moore, L.R.: A statistical evaluation of multiplicative congruent random number generators with modulus 2^{31} -1. J. Am. Stat. Assoc. **77**, 129–136 (1982)

Gentle, J.E.: Cholesky Factorization. In: Numerical Linear Algebra for Applications in Statistics, pp. 93–95. Springer, Berlin (1958)

Hahn, G.J., Shapiro, S.S.: Statistical Models in Engineering. Wiley, New York (1994)

Hastings Jr., C.: Approximation for Digital Computers. Princeton University Press, Princeton (1955)

Hines, W.W., Montgomery, D.C., Goldsman, D.M., Borror, C.M.: Probability and Statistics in Engineering. Wiley, New Jersey (2003)

Kelton, W.D., Sadowski, R.P., Sturrock, D.T.: Simulation with Arena. McGraw Hill, New York (2007)

Law, A.M.: Simulation Modeling and Analysis, 4th edn. McGraw Hill, Boston (2007)

Lehmer, D.H.: Mathematical methods in large scale computing units. Ann. Comput. Lab. **26**, 142–146 (1951). Harvard University

Lewis, P.S.W., Goodman, A.S., Miller, J.M.: A pseudo random number generator for the system/360. IBM Syst. J. **8**, 136–146 (1969)

Payne, W.H., Rabung, J.R., Bogyo, T.P.: Coding the lehmer pseudorandom number generator. Commun. Assoc. Comput. Mach. **12**, 85–86 (1969)

Rand Corporation: A Million Random Digits with 100,000 Normal Deviates, Santa Monica (1946)

Rose, C., Smith, M.D.: Order statistics. Math. Stat. Math. **9.4**, 311–332 (2002)

Schever, E.M., Stoller, D.S.: On the generation of normal random vectors. Technometrics **4**, 278–281 (1962)

Wilk, M.B., Gnanadesikan, R.: Probability plotting methods for the analysis of data. Biometrika **55**, 1–17 (1968)

Zanakis, S.H.: A simulation study of some simple estimators for the three-paramater Weibull distribtuion. J. Stat. Comput. Simul. **9**, 101–116 (1979)

N.T. Thomopoulos, *Essentials of Monte Carlo Simulation: Statistical Methods for Building Simulation Models*, DOI 10.1007/978-1-4614-6022-0,
© Springer Science+Business Media New York 2013

Index

A

Accept-reject (AR) method, 5, 15, 17–18, 31, 32
Active redundancy, 6, 71, 73–74, 78
Approximation formula, 5, 27, 41, 44, 96
Arrival rate, 54, 72, 80, 83, 91
Assembly line, 80, 100, 108
Autocorrelation, 13, 114–115
Average, 4, 6, 7, 12, 54, 55, 71–73, 81, 82, 85–89, 91, 92, 101, 109, 110, 112, 114–117, 119, 121–126, 128, 130, 131, 137

B

Basic, 2
Basic fundamentals, 3–4
Batch arrivals, 6, 71, 73, 78
Batch size, 73, 82
Bathtub, 34, 118, 139
Bernoulli, 5, 45, 47–49, 55
Beta, 5, 7, 27, 33–35, 44, 116–118, 123, 137–140
Beta function, 34
Binomial, 5, 45, 48–49, 52, 55, 60, 116, 119, 124, 125, 133, 134
32-bit word length, 11–12, 14
Bivariate lognormal, 5, 47, 65, 66, 70
Bivariate normal, 5, 57, 62, 64–65, 68, 70
Buffer, 82–84

C

C++, 2
Candidate, 7, 113, 116, 117, 119
Central limit theorem, 30, 38, 41, 94–95, 104
Chi square, 5, 12–13, 27, 40–44

Chi-square variate, 41–42
Cholesky decomposition, 57, 66–68, 70
COBOL, 2
Coefficient of variation (COV), 85–89, 115–118
Coefficients, 84, 124
Collecting the data, 114
Comparing p_1 and p_2, 105–106
Comparing two options, 100–101
Comparing x_1 and x_2, 101
Component, 73–76, 99
Composition, 21, 24, 25
Computer languages, 2
Computer simulation, 2, 6, 14, 26, 78–80, 83, 90, 91, 135, 145
Computer simulation software, 3
Conditional, 63, 64, 66
Conditional distributions, 63–64
Confidence interval, 4, 6, 91–98, 100–107, 111, 112
Confidence limits, 93–95, 97, 98, 100–103, 105–107, 111
Congestion time, 100, 101, 108
Congruent, 10–11, 14
Constant, 6, 53, 71, 72, 78
Continuous, 2–5, 7, 15–22, 24, 27–44, 57, 113, 116, 126, 132, 135, 137, 139
Continuous distributions, 22, 78, 117, 120, 127, 128
Continuous uniform, 5–7, 11, 12, 26–31, 37, 38, 44, 46–51, 55, 58, 73, 75, 91, 92, 116, 117, 120, 121, 127, 129, 137–138
Continuous variable, 16, 139
Control variable, 6, 7, 91, 100, 113
Convolution method, 38
Correlation, 63, 65
Covariance, 63, 66–68

N.T. Thomopoulos, *Essentials of Monte Carlo Simulation: Statistical Methods for Building Simulation Models*, DOI 10.1007/978-1-4614-6022-0, © Springer Science+Business Media New York 2013

Printed by Publishers' Graphics LLC
CAMZ131201.20.07.170